Author

Anton Good
Forschungsinstitut für Mathematik, ETH Zürich
8092 Zürich, Switzerland

AMS Subject Classifications (1980): 10D15, 10G10, 22E40, 30F35, 33A75, 43A80

ISBN 3-540-12713-5 Springer-Verlag Berlin Heidelberg New York Tokyo
ISBN 0-387-12713-5 Springer-Verlag New York Heidelberg Berlin Tokyo

Printing and binding: Beltz Offsetdruck, Hemsbach/Bergstr.
2146/3140-543210

Table of contents

1. Introduction

In number theory identities and functional equations have played an eminent role for a long time. They often come in pairs as e.g. the 1-dimensional Poisson summation formula and the functional equation for the Riemann zeta-function. With their help many arithmetical functions have been studied. In these notes we derive a large number of identities from our investigations of certain meromorphic functions which satisfy functional equations relating their values at s and $1-s$. In very special cases these functions are essentially Hecke's zeta-functions with Grössencharacters of quadratic fields. In general, the functional equations do no longer follow from a Poisson summation formula but basically stem from the fact that selfadjoint operators have only real eigenvalues. Our results have links with problems from different branches of mathematics. We mention just one example from the following fields:

algebra: structure of discrete groups.

analysis: spectral theory of Laplacians.

geometry: distribution of geodesics on Riemann surfaces (cf. Example 3 in sect. 11).

number theory: Mean-value theorems for zeta-functions (cf. [4],[8],[11]).

automorphic forms: Rankin-Selberg convolution method for non-arithmetic groups (cf. [7]).

The identities we are going to prove connect algebraic data of discrete groups with spectral data of differential operators. The general set up can be described as follows. Let S be a manifold on which there is given a differential operator D and a measure ω.

Assume that D and ω are invariant under the action of a Lie group G on S. If Γ is a discrete subgroup of G then D and ω have natural projections to the orbit space $\Gamma\backslash S$. It often happens that in this way D defines a self-adjoint operator in $L_2(\Gamma\backslash S)$, the Hilbert space of ω-square integrable functions on $\Gamma\backslash S$. Under these assumptions, it is not surprising that relations should exist between the spectral theory of D in $L_2(\Gamma\backslash S)$ and the algebraic structure of Γ. A priori, however, it is neither clear how such relations will look like nor obvious whether they can be listed completely.

If G is abelian and non-compact such relations are provided by the Poisson summation formula. For non-abelian G it was Selberg [20] who first established such a connection in his celebrated trace formula. This formula relates the eigenvalues of D in $L_2(\Gamma\backslash S)$ with the conjugacy classes of Γ. Selberg gave a definitive form of the trace formula in the fundamental case when G is the group of real unimodular 2×2 matrices, Γ a finitely generated Fuchsian group of the first kind, S the upper half-plane, D the hyperbolic Laplacian and ω the hyperbolic measure on S. In general, the presence of continuous spectra and the occurring integral transforms create great problems which are not yet fully overcome.

It is possible to relate not only global data like eigenvalues and conjugacy classes but also data of a local nature, e.g. the <u>values</u> of eigenfunctions and the <u>stabilizers</u> at two given points. The main goal of these notes is to establish links of such a local nature. We restrict ourselves to the fundamental case mentioned in the preceding paragraph. There, however, we are able to give a <u>complete</u> list of those relations stemming from <u>Fourier series</u> expansions of automorphic functions.

We recall that Fourier series expansions of automorphic functions always occur in connection with stabilizers Γ_ξ in Γ, where one distinguishes the following three types:

(i) ξ is any point in the upper half-plane (elliptic).

(ii) ξ is a cusp of Γ (parabolic).

(iii) ξ is a pair of hyperbolic fixpoints for Γ (hyperbolic).

Since the action of Γ_ξ on the upper half-plane is always cyclic a Fourier series expansion exists for every Γ_ξ-invariant function. One can therefore attach Fourier coefficients to automorphic functions at every ξ.

From an analytical point of view, the main interest is in the spectral decomposition of D in $L_2(\Gamma\backslash S)$. Here the methods of functional analysis work very well on an abstract level (cf. [19]). By themselves, however, they yield little on more concrete questions as e.g. the rate of growth of the spectrum. We therefore formulate problem

(I) Analyze the eigenfunctions occurring in the spectral decomposition of D in $L_2(\Gamma\backslash S)$ through their various Fourier coefficients.

On the other hand, the main algebraic concern is in the structure of the discrete group Γ. We aim at a better understanding of Γ by reducing Γ modulo its abelian subgroups. This leads to problem

(II) Determine the structure of the double coset space $\Gamma_\xi\backslash\Gamma/\Gamma_\chi$ for an arbitrary pair (Γ_ξ,Γ_χ) of stabilizers.

Although (I) and (II) seem to be rather different problems there is a kind of duality between them. This duality is the heart of the matter. Here we shall unravel it by establishing numerous identities connecting (I) with (II). The effect will be that progress with (I) usually yields new insights into (II) and vice versa.

There has been previous work on some special cases of the general program outlined above. First of all, Selberg derived his trace formula by integrating an identity for a weighted spectral kernel function over the diagonal. The identity he started from is one of the infinitely many identities we shall obtain if in (II) both ξ and χ are of the elliptic type (cf. Corollary 2 in sect. 10). It was this example that prompted our title. In recent years the case when both ξ and χ are of the parabolic type has been studied extensively in [1], [2], [13] and [18]. We supplement their work by adding further identities in this case (cf. Corollary 1 in sect. 10). It should be noted that these recent papers were anticipated by Selberg [21] who did not prove the identities but set up the basic analytical machinery.

So far, those identities have mainly been used to investigate all kinds of averages over the spectrum or over the double cosets. Here we only present such applications which straightforwardly follow from taking trivial bounds after the use of a single identity (cf. Theorem 2 and Theorem 4). For this type of result one could get around with much less explicit connections between spectrum and group (cf. [6], [10]). In more complex applications, however, infinitely many identities enter the game. After their use further cancellations are obtained from inequalities of the large sieve type (cf. [3], [9]) or from functional equations for Dirichlet series (cf. [7]). Moreover uniformity in parameters then becomes important. It is this second type of applications for which the identities prove to be particularly useful or even indispensible.

The Fourier expansions of type (i)-(iii) were first studied systematically by Petersson [17] in case of the holomorphic automorphic forms. He introduced three classes of Poincaré series and computed their Petersson inner product with an arbitrary cusp form. Petersson treated one class after the other. If we proceeded likewise, we would have to consider problem (II) nine times over. For obvious reasons one tries to avoid this by looking for a unified approach. Our special attention is indeed given to a simultaneous treatment of all cases in problem (II). We make it evident that the main results always follow from the same algebraic manipulations, at least if all questions of convergence are ignored. Some differences do remain, however. They are mostly analytical in nature and due to a changing pattern of convergence. Although different functions have to be studied in the different cases, their behaviour is very similar as one of the parameters gets large. This explains why in all cases the asymptotic results look the same (cf. Theorem 2 and Theorem 4). Many problems which have received separate attention in the literature are subsumed in the basic problem (II). Therefore our approach puts on equal footing so different things as the hyperbolic lattice point problem, the estimation of sums of Kloosterman sums or the distribution of the geodesics in covering surfaces lying over a fixed closed geodesic (cf. the three examples at the end of sect. 11). Moreover, cases considered here for the first time happen to provide new insights into old number-theoretical problems, as the following example shows.

The 2-dimensional Poisson summation formula provides a good attack on the classical lattice point problem in Euclidean circles. In particular, it proves the uniform distribution modulo 1 of the numbers $\frac{1}{2\pi}\arg w$ when w runs through the Gaussian integers according to any ordering compatible with their norms. By the Moebius inversion formula this continues to hold if one admits only primitive w, i.e. w with coprime real and imaginary parts. On the other hand, the restriction to primitive lattice points allows to refine the distribution problem. For then there exists a w^* such that w and w^* form an integral basis for the Gaussian integers. Since the numbers w^* are not uniquely determined it is not sensible to study the distribution of their arguments. One notice, however, that $\frac{w^*}{w}$ is uniquely determined modulo 1 by w and the inequality $\mathrm{Im}\,\frac{w^*}{w} > 0$. Under the latter condition the pairs $(\frac{1}{2\pi}\arg w, \mathrm{Re}\,\frac{w^*}{w})$ prove to be uniformly distributed in the unit square, when w runs through the primitive Gaussian integers ordered as before. This result can no longer be deduced from the Poisson summation formula but requires the use of harmonic analysis on the modular curve. Indeed it is a consequence of our corollary in sect. 11 with ξ of the elliptic and χ of the parabolic type.

Our unified approach to problem (II) did not evolve from an adaptation of known proofs for special cases. On the contrary, it is just to say that we also provide new proofs in the cases previously considered. As a guide-line, problem (II) helped us immensely to choose adequate parameters and definitions, but it also led us to break with some traditions. According to the general set up we split our arguments into an algebraic and an analytical part.

On the algebraic side we require a number of double coset decompositions of G among which one does not only find the Iwasawa or Bruhat decompositions but also some new. We give these decompositions in terms of maps from G to its Lie algebra while they are usually defined by maps in the opposite direction. There are two reasons for it. First of all we start with a discrete subgroup Γ whose elements we want to parametrize. Our decompositions then allow us to define generalized Kloosterman sums in Lie algebraic terms. These sums describe the discrete content of the group-theoretical side in our identities. Secondly, our maps are given so explicitly that we are able to write down the functions which govern the effect of the regular representation on periodic eigenfunctions. These functions describe the continuous content of the group-theoretical side in our identities. The whole situation reminds one of the Hardy-Littlewood method, where the main parts split into the singular series and the singular integral.

On the analytical side progress comes from the systematic use of eigenfunctions, integral representations and functional equations. We follow Selberg [20] by deducing everything from the basic eigenfunctions

$$z \longmapsto (\mathrm{Im} z)^s .$$

We do no longer discuss the periodic eigenfunctions in terms of solutions to ordinary differential equations by separating variables. Instead, we directly define them by integrals involving those basic eigenfunctions. Our definitions yield more coherent asymptotic results on Fourier coefficients than the usual normalizations. More importantly, however, they directly lead to integral representations for the functions alluded to in the preceding paragraph. Since it proves to be irrelevant to identify the particular functions represented by these integrals, no books on special functions have to be consulted. The Poincaré series we consider here are eigenfunctions as well. To establish their analytic continuation we prefer to work with truncations rather than the usual first order approximation to power series expansions (cf. [15], [21]). Being eigenfunctions, our Poincaré series satisfy a functional equation and have much simpler Fourier series expansions than e.g. the $U_m(z,s)$ in [21]. However, the functions we are really after are those corresponding to the logarithmic derivative of the Selberg zeta-functions. These now occur as Fourier coefficients of our Poincaré series from which their own analytic continuation and functional equation are derived. After that it is a rather simple exercise in contour integration to deduce the desired identities. On their analytical side, the contribution of the discrete spectrum comes from the residues and that of the continuous spectrum comes from the functional equation of those functions.

On a purely formal level, the basic steps in our procedure are not very difficult to carry out. It requires some efforts, however, to make them rigorous since technical complications do arise. Everything works out as expected if both ξ and χ are of the elliptic type. If one of them is parabolic, the straightforward approach leads to some merely conditionally convergent integrals. We then have to justify their interchanges by suitable integrations by parts. The most complicated situations occur if hyperbolic types are involved. On one hand, we split functions into even and odd parts in order to obtain eigenfunctions which are periodic with respect to a hyperbolic 1-parameter subgroup and which decay rapidly enough near the boundary of the upper half-plane. On the other hand, our decompositions of G contain two parities which can be non-trivial only with hyperbolic types. They are responsible that up to four terms may occur where, naively, one would expect just one. Moreover, there is a middle part (cf. (3.11)) in our decompositions which is non-empty again only with hyperbolic types. It gives rises to a finite number of additional terms in the identities.

We develop ab ovo all parts of our method for which there are no direct references available. Only in Corollary 1 and the remark following it (see sect. 10) we depart from our strategy by identifying certain integrals with Bessel functions. This is to show that Corollary 1 contains the result we stated and used in [7], [8], [9]. It is perhaps worthwhile to note that numerous results on special functions (especially but not only on Bessel and Legendre functions) drop out as a by-product of our considerations if one identifies the integrals systematically. Moreover, large parts of [5], [15] and [16] are then seen to be subsumed in our statements on Poincaré series.

We close the introduction by outlining the content of the different sections.

Section 2: We set up notation and summarize familar facts on the spectral resolution of the hyperbolic Laplacian. Later on, we shall particularly use the expansion of automorphic functions in terms of automorphic eigenfunctions (cf. (2.12), (2.13)).

Section 3: First we split G into three parts (cf. (3.11)) by means of fixpoint properties (cf. (3.9), (3.10)). We introduce the functions ${}_{\xi}\Lambda_{\chi}^{\ell}$ and ${}_{\xi}\Lambda_{\chi}^{r}$ (cf. (3.13)-(3.15)) which obey simple transformation laws. These functions will play a crucial rôle in our double coset decompositions. From an analytical point of view we have some freedom in choosing the parameters. Our choices, however, yield decompositions of G (Lemma 1) and of its complexification (Lemma 2) which are unique in a geometrical sense (cf. the remark following the proof of Lemma 1). Although we are primarily interested in G and its subgroups, the complexification has also to be considered for reasons which become clear only in sect. 4 and 6.

Section 4: We define the periodic eigenfunctions U_{ξ} and V_{ξ} (cf.(4.7), (4.8)) by integrals over group translates of the basic eigenfunctions mentioned above. Since the integration stays inside G, the U_{ξ} are smooth in the upper half-plane but do not decay near the boundary. On the other hand, the V_{ξ} decay. In order to achieve this we had to sacrifice smoothness by moving contours into the complexification. For several reasons we determine the boundary behaviour of U_ξ and V_ξ. This requires the complete evaluation of certain integrals which can always be reduced to beta- or gamma-integrals. We essentially characterize U_ξ and V_ξ by intrinsic properties (Lemma 3). This characterization and the asymptotic behaviour then imply functional equations for U_ξ and V_ξ (Lemma 4). They determine the functions γ_ξ which also show up in all subsequent functional equations. Here as well as later on one observes degeneracies in the functional equations if basic eigenfunctions are involved.

Section 5: After a reduction to a normal form we study the Fourier series expansions at ξ of square integrable eigenfunctions (Lemma 5). In these expansions only the U_ξ can turn up. We define the generalized Kloosterman sums (cf. (5.10), (5.11)) by means of ${}_\xi\Lambda_\chi^\ell$ and ${}_\xi\Lambda_\chi^r$. If Γ is the modular group and $\xi = \chi = \infty$ (i.e. parabolic), these sums are in fact ordinary Kloosterman sums (cf. Remark 1 after Lemma 6). We also show that certain subsets of G can contain no or only finitely many elements of a discrete subgroup Γ (Lemma 6).

Section 6: This is the most technical section. Basically, we have to justify the interchange of integration in double integrals. We introduce functions ${}_\xi J_\chi$ for which integral representations are derived (Lemma 7). To a large extent, the effect of the regular representation on V_ξ can be described by means of ${}_\xi J_\chi$ (cf. the remark at the end of sect. 6). In some cases which play a special rôle later on, ${}_\xi J_\chi$ is given by a rather simple expression (Lemma 8). Moreover, all ${}_\xi J_\chi$ approach that expression as one of the parameters gets large (Lemma 9). In Lemma 10 we give integral representations for ${}_\xi I_\chi$ which are related to U_ξ in the same way as the ${}_\xi J_\chi$ to V_ξ. There we also prove functional equations for ${}_\xi J_\chi$ and ${}_\xi I_\chi$. In Lemma 11 we obtain similar results for functions ${}_\xi i_\chi$ which arise from the middle part (cf. (3.11)) of our decompositions.

Section 7: By means of V_ξ we introduce the Poincaré series P_ξ (cf. (7.1)) among which one finds e.g. the Eisenstein series and the resolvent kernel. We establish convergence for the truncated P_ξ^ψ (cf. (7.2)) in Lemma 12. In (7.11) and (7.12) we define functions ${}_\xi p_\chi$ and ${}_\xi P_\chi$ in terms of generalized Kloosterman sums and ${}_\xi i_\chi$ or ${}_\xi J_\chi$ respectively. These functions appear in the Fourier series expansion of P_ξ at χ (Proposition 1). Finally, special attention is given to the Fourier coefficients of Eisenstein series (cf. (7.18)).

Section 8: In preparation for the following section we study the functions Ψ_ξ and ψ_ξ (cf. (8.1), (8.3)). They are related in Lemma 13, where we determine special values of ψ_ξ as well. These values turn up in residues later on. We evaluate an integral which is an untruncated version of the particular Ψ_ξ bound up with the resolvent kernel (Lemma 14).

Section 9: The expansion of the P_ξ^ψ into automorphic eigenfunctions yields their analytic continuation (Proposition 2(i)). Here Ψ_ξ turns up and the previous two lemmas imply convergence. We determine principal parts (Proposition 2(ii)) and functional equations (Proposition 2(iii)) for the Poincaré series P_ξ. For the latter a symmetry has to be used which generalizes that of the so called constant term matrix for the Eisenstein series (cf. (9.10) and the subsequent Remark 1). Analytic continuation, principal parts and functional equations for ${}_\xi p_\chi$ and ${}_\xi P_\chi$ are given in Proposition 3(i)-(iii). One notices that the ${}_\xi P_\chi$ and the logarithmic derivative of Selberg zeta-functions have similar analytic properties. For later use we also give many functional equations for the Fourier coefficients of Eisenstein series (Proposition 3 (iv)). Finally we crudely bound the growth of ${}_\xi P_\chi$ in vertical strips (Lemma 15).

Section 10: We give a first form of the identities which are sum formulae of the Bruggeman type [1]. We distinguish the 'generic' case (Theorem 1 (i)) and a degenerate situation (Theorem 1 (ii)). The asymmetry in the weight function h can be removed if one wants to (cf. the remark after Theorem 1). The integral transforms ${}_{\xi}h_{\chi}$ and ${}_{\xi}H_{\chi}$ of h correspond to the so called Selberg transform. At first they are given in terms of ${}_{\xi}i_{\chi}$ and ${}_{\xi}I_{\chi}$ respectively. We now write them as compositions of two transforms of which the inner always turns out to be an ordinary Fourier transformation. We restate Theorem 1 for the purely parabolic case (Corollary 1) and the purely elliptic case (Corollary 2) in more conventional terms. These cases cover [1] and the identities for point pair invariants in [20]. Theorem 1 is very useful in the study of averages over the spectrum. We investigate the square mean of Fourier coefficients (Theorem 2). Thanks to our normalizations its growth is essentially the same in all cases.

Section 11: We derive a second form of the identities which is useful to study averages over double cosets (Theorem 3). In principle, this requires the inversion of the integral transforms in Theorem 1. For the purely parabolic case, Fourier coefficients of holomorphic automorphic forms came up in this inversion (cf. [2], [3]). We can get by without them on making only an approximate inversion by means of our functional equations. The essential terms on the right hand sides in Theorem 3 only involve Mellin transformations, while the ugly parts are much smaller asymptotically (Theorem 4). In this last theorem we show that great cancellations always occur in sums of generalized Kloosterman sums. In particular, it yields a 2-dimensional uniform distribution law for the elements in $\Gamma_{\xi}\backslash\Gamma/\Gamma_{\chi}$ (Corollary to Theorem 4). Our remainder term is as strong as the usual bound for the hyperbolic lattice point problem (Example 1). It falls short of Kuznietsov's bound for ordinary Kloosterman sums [13] since, in general, no A. Weil's estimate is available (Example 2). Finally, we consider an application to the distribution of geodesics (Example 3).

2. Preliminaries

Let $\mathbb{Z}$ denote the integers, $\mathbb{R}$ the real and $\mathbb{C}$ the complex numbers. The group $SL_2(\mathbb{C})$ of 2×2-matrices M with complex entries and determinant 1 acts on points z of the Riemann sphere $\bar{\mathbb{C}} = \mathbb{C} \cup \{\infty\}$ by fractional linear transformations:

$$(M,z) \longmapsto M(z) = \frac{az+b}{cz+d} \quad , \quad \text{where} \quad M = \begin{pmatrix} a & b \\ c & d \end{pmatrix} . \tag{2.1}$$

We shall often write z_M instead of $M(z)$. Here we are mainly dealing with the group $G = SL_2(\mathbb{R})$ consisting of all matrices in $SL_2(\mathbb{C})$ with real entries. The action of G on the Riemann sphere leaves both the upper half-plane $\mathfrak{H}$ and its boundary $\mathbb{R} \cup \{\infty\}$ invariant. Moreover there exist a second order linear differential operator Δ and a measure ω on $\mathfrak{H}$ which are G-invariant. We introduce them in terms of the standard global coordinates on $\mathfrak{H}$: $z = x+iy \longmapsto (x,y)$, where x,y are real and $y > 0$. Then Δ -the so-called Laplacian - is given by

$$\Delta = y^2 \left(\frac{\partial^2}{\partial x^2} + \frac{\partial^2}{\partial y^2} \right) , \tag{2.2}$$

while on a Lebesgue measurable set $\mathfrak{U}$ in $\mathfrak{H}$

$$\omega(\mathfrak{U}) = \int_{\mathfrak{U}} d\omega(z) = \int_{\mathfrak{U}} \frac{dxdy}{y^2} , \tag{2.3}$$

where $dxdy$ stands for integration with respect to the Lebesgue measure in the (x,y)-plane. The G-invariance characterizes Δ and ω up to scalar multiples.

Now let Γ be a finitely generated Fuchsian group of the first kind. Equivalently we may say that Γ is a discrete subgroup of G having a fundamental domain $\mathfrak{F}$ of finite ω-measure. In the following we always denote points in $\mathfrak{H}$ by ζ, ζ' and cusps of Γ by ϑ, ϑ' , i.e. ϑ, ϑ' are fixpoints of parabolic elements in Γ. Moreover η, η' always stand for ordered triples $(\eta_1, \eta_2, \varepsilon)$, $(\eta_1', \eta_2', \varepsilon')$, where η_1, η_2 and η_1', η_2' respectively are the fixpoints of hyperbolic elements in Γ and $\varepsilon, \varepsilon' = \pm 1$. For the next section the third components of η, η' are of no significance. Their rôle will become clear only later on. They describe a parity we were led to introduce in order to ensure certain Poincaré series from divergence. If $\eta = (\eta_1, \eta_2, \varepsilon)$ and M is in $SL_2(\mathbb{C})$ we put

$$\eta^* = (\eta_2, \eta_1, \varepsilon) \tag{2.4}$$

and

$$M(\eta) = (M(\eta_1), M(\eta_2), \varepsilon) \; . \tag{2.5}$$

The letter ξ will always refer to a ζ, ϑ or η while χ will denote a ζ', ϑ' or η'. Thus Γ acts on all these letters. Consequently the following two equivalence relations can be introduced: By definition

$$\xi \sim \chi \text{ means that } \begin{cases} \xi = M(\chi) \; , & \text{if } \xi = \zeta, \vartheta \; , \\ (\eta_1, \eta_2, \pm 1) = M(\chi) \; , & \text{if } \xi = (\eta_1, \eta_2, \varepsilon) \; , \end{cases} \tag{2.6}$$

and

$$\xi \approx \chi \quad \text{means that} \quad \xi = M(\chi) \tag{2.7}$$

with a suitably chosen M in Γ and a suitable sign in the second line of (2.6).

Let $L_2(\Gamma \backslash \mathfrak{H})$ denote the space of Γ-invariant function on $\mathfrak{H}$ which are ω-square-integrable on the fundamental domain $\mathfrak{F}$. In $L_2(\Gamma \backslash \mathfrak{H})$ we define a scalar product by

$$\langle f, g \rangle = \int_{\mathfrak{F}} f(z) \bar{g}(z) \, d\omega(z) \tag{2.8}$$

where $\bar{g}$ denotes the complex conjugate of g. Since ω is G-invariant $\langle f, g \rangle$ does not depend on the chosen $\mathfrak{F}$. Modulo identification of functions being equal ω-almost everywhere $\langle , \rangle$ turns

$L_2(\Gamma \backslash \mathfrak{H})$ into a Hilbert space.

By the G-invariance the Laplacian Δ carries Γ-invariant C^∞-functions into Γ-invariant C^∞-functions. This action extends to a self-adjoint operator in $L_2(\Gamma \backslash \mathfrak{H})$ which we call the Laplacian in $L_2(\Gamma \backslash \mathfrak{H})$. Next we summarize some basic properties of its spectral decomposition. Proofs of these facts can be found in [12], [19].

The Laplacian in $L_2(\Gamma \backslash \mathfrak{H})$ always has a non-empty discrete spectrum. Its contribution to the spectral resolution can be described in terms of a maximal system $(e_j)_{j \geq 0}$ of orthonormal functions e_j in $L_2(\Gamma \backslash \mathfrak{H})$ for which there is a complex number s_j with

$$\Delta e_j(z) + s_j(1-s_j)e_j(z) = 0 , \quad z \text{ in } \mathfrak{H} . \tag{2.9}$$

Since $-\Delta$ is non-negative we may and will assume that either $\frac{1}{2} \leq s_j \leq 1$ or $s_j = \frac{1}{2} + it_j$ with $t_j > 0$. In particular, we always take $s_0 = 1$, whence e_0 is a constant function, namely

$$e_0(z) = (\omega(\mathfrak{F}))^{-1/2} . \tag{2.10}$$

The Laplacian in $L_2(\Gamma \backslash \mathfrak{H})$ has a continuous spectrum if and only if Γ contains parabolic elements. The contribution of the continuous spectrum to the spectral resolution can be described in terms of finitely many Eisenstein series $E_\iota(z,s)$, $\iota = 1,\ldots,\kappa$. As functions of the first variable the E_ι are Γ-invariant eigenfunctions of Δ, namely

$$\Delta E_\iota(z,s) + s(1-s)E_\iota(z,s) = 0 , \quad z \text{ in } \mathfrak{H} . \tag{2.11}$$

These Eisenstein series are among the Poincaré series we consider in section 7. As such they are analytic functions of $s = \sigma + it$ in the half-plane $\sigma > 1$. A. Selberg showed that the $E_\iota(z,s)$ extend to meromorphic functions on the s-plane and that they have no poles on the line $\sigma = 1/2$.

The spectral resolution of the Laplacian in $L_2(\Gamma\backslash\mathfrak{H})$ can now be expressed by means of the eigenfunctions in (2.9) and (2.11). Actually we shall have to decompose only bounded C^∞-functions in $L_2(\Gamma\backslash\mathfrak{H})$. For such functions f the expansion into eigenfunctions takes the more convenient form

$$f(z) = \sum_{j\geq o} <f,e_j>e_j(z) + \frac{1}{4\pi i} \sum_{\iota=1}^{\kappa} \int_{(1/2)} <f,E_\iota(\cdot,s)>E_\iota(z,s)\,ds \,, \qquad (2.12)$$

where $\int_{(\varrho)}$ denotes integration along the line $\sigma = \varrho$ from $\varrho - i\infty$ to $\varrho + i\infty$ and the right hand side converges absolutely and locally uniformly on $\mathfrak{H}$. In (2.12) $<f,E_\iota(\cdot,s)>$ is also given by an absolutely convergent integral although $E_\iota(z,s)$ is not in $L_2(\Gamma\backslash\mathfrak{H})$ for s on $\sigma = 1/2$ except possibly for $s = 1/2$. Thus the integral term in (2.12) is not meaningful for all f in $L_2(\Gamma\backslash\mathfrak{H})$ and indeed the expansion of an arbitrary f into eigenfunctions with convergence in the L_2-norm differs somewhat from (2.12). We do not reproduce it here since (2.12) suffices for our purposes.

Finally we recall the completeness relation

$$<f,f> = \sum_{j\geq o} |<f,e_j>|^2 + \frac{1}{4\pi i} \sum_{\iota=1}^{\kappa} \int_{(1/2)} |<f,E_\iota(\cdot,s)>|^2 ds \qquad (2.13)$$

for f in $L_2(\Gamma\backslash\mathfrak{H})$ provided that $<|f|,|E_\iota(\cdot,s)|> < \infty$ on $\sigma = 1/2$ and $\iota = 1,\dots,\kappa$.

3. Decompositions of G

In the notation of (2.1) the elements of G are in 1-1-correspondence with the points on the 3-dimensional hypersurface $ad - bc = 1$ in $\mathbb{R}^4$. Unfortunately this fact does not reveal very much about G as a group. In this section we discuss various parametrizations of G which are especially tailored to its group structure.

First let

$$X_\zeta = \begin{pmatrix} 0 & 1/2 \\ -1/2 & 0 \end{pmatrix}, \quad X_\vartheta = \begin{pmatrix} 0 & 1 \\ 0 & 0 \end{pmatrix} \quad \text{and} \quad X_\eta = \begin{pmatrix} 1/2 & 0 \\ 0 & -1/2 \end{pmatrix}. \tag{3.1}$$

Thus if exp denotes the exponential map from the complexified Lie algebra of G to $SL_2(\mathbb{C})$ we have for τ in $\mathbb{C}$

$$\exp(\tau X_\zeta) = \begin{pmatrix} \cos\frac{\tau}{2} & \sin\frac{\tau}{2} \\ -\sin\frac{\tau}{2} & \cos\frac{\tau}{2} \end{pmatrix}, \quad \exp(\tau X_\vartheta) = \begin{pmatrix} 1 & \tau \\ 0 & 1 \end{pmatrix}$$
$$\exp(\tau X_\eta) = \begin{pmatrix} e^{\tau/2} & 0 \\ 0 & e^{-\tau/2} \end{pmatrix}. \tag{3.2}$$

We recall that every non-trivial 1-parameter subgroup of G is conjugate to a

$$H_\xi = \{\exp(\tau X_\xi) \mid \tau \text{ in } \mathbb{R}\}. \tag{3.3}$$

Our parametrizations of G are built up from the double cosets $H_\xi M H_\chi$, M in G. Since H_ξ and H_χ are 1-dimensional the set of these double cosets is essentially 1-dimensional as well. To make a precise statement we require further definitions.

Let $\mathfrak{f}_\xi$ denote the set of fixpoints of $\exp(X_\xi)$, i.e.

$$\mathfrak{f}_\xi = \begin{cases} \{\pm i\}, & \text{if } \xi = \zeta, \\ \{\infty\}, & \text{if } \xi = \vartheta, \\ \{0,\infty\}, & \text{if } \xi = \eta, \end{cases} \tag{3.4}$$

and

$$\mathfrak{F}_\xi = \begin{cases} \mathfrak{f}_\xi, & \text{if } \xi = \zeta \text{ or } \vartheta, \\ \mathfrak{f}_\xi \cup \{iy \mid y>0\}, & \text{if } \xi = \eta. \end{cases} \tag{3.5}$$

Whenever logarithms occur hereafter the following convention applies. For z in $\bar{\mathbb{C}} - \mathfrak{f}_\eta$, $\arg z = \mathrm{Im}(\log z)$ lies in the interval $(-\pi, \pi]$ unless it is stated otherwise.

On $\bar{\mathbb{C}} - \mathfrak{f}_\xi$ we define maps w_ξ by

$$w_\xi(z) = \begin{cases} \frac{1}{i}\log\frac{z-i}{z+i} & , \text{ if } \quad \xi = \zeta \quad , \\ z & , \text{ if } \quad \xi = \vartheta \quad , \\ \log z & , \text{ if } \quad \xi = \eta \quad . \end{cases} \tag{3.6}$$

Thus $w_\xi(z) = u_\xi(z) + iv_\xi(z)$, where

$$u_\xi(z) = \begin{cases} \arg\left(\frac{z-i}{z+i}\right) & , \text{ if } \quad \xi = \zeta \quad , \\ \operatorname{Re} z & , \text{ if } \quad \xi = \vartheta \quad , \\ \log|z| & , \text{ if } \quad \xi = \eta \quad , \end{cases}$$
$$v_\xi(z) = \begin{cases} -\log\left|\frac{z-i}{z+i}\right| & , \text{ if } \quad \xi = \zeta \quad , \\ \operatorname{Im} z & , \text{ if } \quad \xi = \vartheta \quad , \\ \arg z & , \text{ if } \quad \xi = \eta \quad . \end{cases} \tag{3.7}$$

It follows from (3.2), (3.6) and (3.7) that for $M = \exp(\tau X_\xi)$

$$\begin{aligned} w_\xi(z_M) &\equiv \tau + w_\xi(z) \qquad (\text{mod } \ell_\xi) , \quad \tau \text{ in } \mathbb{C} , \\ u_\xi(z_M) &\equiv \tau + u_\xi(z) \qquad (\text{mod Re } \ell_\xi) , \quad \tau \text{ in } \mathbb{R} , \end{aligned} \tag{3.8}$$

with $\ell_\zeta = 2\pi$, $\ell_\vartheta = 0$ and $\ell_\eta = 2\pi i$.

Next let

$${}_\xi\mathcal{I}_\chi = \{M \text{ in } G | M(z) \text{ in } \mathfrak{f}_\xi \text{ for some } z \text{ in } \mathfrak{f}_\chi\} \tag{3.9}$$

and

$${}_\xi\mathcal{J}_\chi = \{M \text{ in } G | M(z) \text{ in } \mathfrak{F}_\xi \text{ for some } z \text{ in } \mathfrak{F}_\chi\} . \tag{3.10}$$

Except for ${}_\eta\mathcal{J}_{\eta'}$ it is straightforward to check that ${}_\xi\mathcal{I}_\chi$ and ${}_\xi\mathcal{J}_\chi$ coincide with the subsets of G given in the notation of (2.1) and (3.3) by the following tables ($\emptyset$ denotes the empty set):

$\xi \backslash \chi$	ζ'	ϑ'	η'
ζ	H_ζ	$\emptyset$	$\emptyset$
ϑ	$\emptyset$	$c=0$	$cd=0$
η	$\emptyset$	$ac=0$	$abcd=0$

${}_\xi\mathcal{I}_\chi$

$\xi \backslash \chi$	ζ'	ϑ'	η'
ζ	H_ζ	$\emptyset$	$H_\zeta H_{\eta'}$
ϑ	$\emptyset$	$c=0$	$cd=0$
η	$H_\eta H_{\zeta'}$	$ac=0$	$\lvert ad\rvert + \lvert bc\rvert = 1$

${}_\xi\mathcal{J}_\chi$

We come back to the entry of $_{\eta}\mathfrak{f}_{\eta'}$ in the proof of Lemma 1 below. By means of the sets introduced G can be written as a disjoint union

$$G = {}_{\xi}G_{\chi} \cup {}_{\xi}g_{\chi} \cup {}_{\xi}\mathfrak{1}_{\chi} \quad , \tag{3.11}$$

where

$${}_{\xi}G_{\chi} = G - {}_{\xi}\mathfrak{f}_{\chi} \quad \text{and} \quad {}_{\xi}g_{\chi} = {}_{\xi}\mathfrak{f}_{\chi} - {}_{\xi}\mathfrak{1}_{\chi} \ . \tag{3.12}$$

Note that each set on the right hand side of (3.11) is left H_{ξ}- and right H_{χ}-invariant.

With $\# \mathfrak{f}_{\chi}$ denoting the cardinality of $\mathfrak{f}_{\chi}$ we now define

$${}_{\xi}\Lambda^{\ell}_{\chi}(M) = \frac{1}{\#\mathfrak{f}_{\chi}} \sum_{z \in \mathfrak{f}_{\chi}} u_{\xi}(z_M) \tag{3.13}$$

unless $\xi = \zeta$ and $\chi = \eta'$. In the latter case we set

$${}_{\zeta}\Lambda^{\ell}_{\eta'}(M) = u_{\zeta}(M(\infty)) + \frac{1}{2} \arg\left(\frac{(M(o)-i)\,(M(\infty)+i)}{(M(o)+i)\,(M(\infty)-i)}\right) \ . \tag{3.14}$$

Note that

$$\frac{1}{\#\mathfrak{f}_{\eta'}} \sum_{z \in \mathfrak{f}_{\eta'}} u_{\zeta}(z_M) = {}_{\zeta}\Lambda^{\ell}_{\eta'}(M) + k\pi \quad ,$$

where $k = 0$ or ± 1 . Definition (3.14) was prompted by geometric considerations. Its advantage over the left hand side of the preceding equation is that its transformation law for the left H_{ζ}-action is defined modulo 2π and not just modulo π . We also introduce

$${}_{\xi}\Lambda^{r}_{\chi}(M) = -{}_{\chi}\Lambda^{\ell}_{\xi}(M^{-1}) \ . \tag{3.15}$$

Finally we set

$$R_{\xi} = \exp(r_{\xi} X_{\zeta}) \quad , \tag{3.16}$$

where $r_{\zeta} = 0$, $r_{\vartheta} = \pi$ and $r_{\eta} = \frac{\pi}{2}$.

We are now ready to state and prove

<u>Lemma 1</u>. (i) For every M in ${}_{\xi}G_{\chi}$ there exist uniquely determined numbers $\delta = {}_{\xi}\delta_{\chi}(M)$, $\delta' = {}_{\xi}\delta'_{\chi}(M)$ and $\nu = {}_{\xi}\nu_{\chi}(M)$ such that

$$M = \pm \exp({}_{\xi}\Lambda^{\ell}_{\chi}(M) X_{\xi}) R_{\vartheta}^{\delta} R_{\xi}^{-1} \begin{pmatrix} \nu & o \\ o & 1/\nu \end{pmatrix} R_{\chi} R_{\vartheta}^{1-\delta'} \exp({}_{\xi}\Lambda^{r}_{\chi}(M) X_{\chi}) \quad ,$$

where $\delta, \delta' = 0$ or 1 and

$$\nu > \begin{cases} 0 \ , & \text{if} \quad \xi = \vartheta \quad \text{or} \quad \chi = \vartheta' \ , \\ 1 \ , & \text{otherwise.} \end{cases}$$

Moreover $\delta = 0$ unless $\xi = \eta$ and $\delta' = 0$ unless $\chi = \eta'$.

(ii) A matrix M belongs to ${}_{\eta}g_{\zeta'}$ or to ${}_{\zeta}g_{\eta'}$ if and only if

$$M = \exp({}_{\eta}\Lambda^{\ell}_{\zeta'}(M)X_{\eta})R_{\eta}\exp({}_{\eta}\Lambda^{r}_{\zeta'}(M)X_{\zeta'})$$

or

$$M = \exp({}_{\zeta}\Lambda^{\ell}_{\eta'}(M)X_{\zeta})R_{\eta}^{-1}\exp({}_{\zeta}\Lambda^{r}_{\eta'}(M)X_{\eta'})$$

respectively.

(iii) For every M in ${}_{\eta}g_{\eta'}$ there is a uniquely determined real number ν with $0 < |\nu-\pi| < \pi$ such that

$$M = \pm\exp({}_{\eta}\Lambda^{\ell}_{\eta'}(M)X_{\eta})\exp(\nu X_{\zeta})\exp({}_{\eta}\Lambda^{r}_{\eta'}(M)X_{\eta'}) \ .$$

<u>Remark</u>. Special cases of Lemma 1 (i) together with (3.9) - (3.12) yield some well-known decompositions of G , e.g. Iwasawa or Bruhat decompositions arise with $(\xi,\chi) = (\zeta,\vartheta')$ or $(\xi,\chi) = (\vartheta,\vartheta')$ respectively.

<u>Proof of Lemma 1</u>: Let

$$N = \exp(-{}_{\xi}\Lambda^{\ell}_{\chi}(M)X_{\xi})M\exp(-{}_{\xi}\Lambda^{r}_{\chi}(M)X_{\chi}) \ . \qquad (3.17)$$

If Lemma 1 (i) holds for M in ${}_{\xi}G_{\chi}$ then by (3.13) - (3.16)

$$N^{-1} = \exp(-{}_{\chi}\Lambda^{\ell}_{\xi}(M^{-1})X_{\chi})M^{-1}\exp(-{}_{\chi}\Lambda^{r}_{\xi}(M^{-1})X_{\xi}) = \mp R_{\vartheta}^{\delta'}R_{\chi}^{-1}\begin{pmatrix} \nu & 0 \\ 0 & 1/\nu \end{pmatrix}R_{\xi}R_{\vartheta}^{1-\delta}$$

since

$$R_{\vartheta}^{-1}\exp(\tau X_{\eta})R_{\vartheta}^{-1} = -\exp(-\tau X_{\eta}) \ .$$

Thus Lemma 1 (i) holds for M in ${}_{\xi}G_{\chi}$ precisely if it holds for M^{-1} in ${}_{\chi}G_{\xi}$.

Now let M belong to $G - {}_{\xi}\mathcal{J}_{\chi}$. Then (3.4), (3.8), (3.13) and (3.14) show that

$${}_{\xi}\Lambda^{\ell}_{\chi}(\exp(\varrho X_{\xi})M\exp(\tau X_{\chi})) \equiv \varrho + {}_{\xi}\Lambda^{\ell}_{\chi}(M) \qquad (\text{mod } \operatorname{Re} \ell_{\xi})$$

for real ϱ and τ . Therefore N defined in (3.17) satisfies

$$ {}_{\xi}\Lambda^{\ell}_{\chi}(N) = 0 \qquad (\text{mod } 2\pi, \text{ if } \xi = \zeta) \tag{3.18}$$

and by (3.15)

$$\begin{aligned} {}_{\chi}\Lambda^{\ell}_{\xi}(N^{-1}) &= {}_{\chi}\Lambda^{\ell}_{\xi}(\exp({}_{\xi}\Lambda^{r}_{\chi}(M)X_{\chi})M^{-1}\exp({}_{\xi}\Lambda^{\ell}_{\chi}(M)X_{\xi})) \\ &= {}_{\xi}\Lambda^{r}_{\chi}(M) + {}_{\chi}\Lambda^{\ell}_{\xi}(M^{-1}) = 0 \qquad (\text{mod } 2\pi, \text{ if } \chi = \zeta') . \end{aligned} \tag{3.19}$$

We readily deduce from (3.4), (3.7), (3.13) and (3.14) that (3.18) holds for N in $G-{}_{\xi}1_{\chi}$ precisely if N fulfills the conditions listed in the following table:

ξ \ χ	ζ'	ϑ'	η'
ζ	$\frac{N(i)}{i}>1$	$N(\infty)=\infty$	$N(o)+N(\infty)=o$, $\lvert N(o)\rvert \geq 1$, $N(o)\neq 1$
ϑ	$\frac{N(i)}{i}>o$	$N(\infty)=o$	$N(o)+N(\infty)=o$
η	$\lvert N(i)\rvert=1$	$\lvert N(\infty)\rvert=1$	$\lvert N(o)N(\infty)\rvert=1$

This table allows us to express (3.18) and (3.19) in terms of the entries of N which we shall denote by the same letters as those of M in (2.1). We proceed by checking one case after the other.

If $(\xi,\chi) = (\zeta,\zeta')$ then (3.18) and (3.19) mean that

$$ai+b = \lambda(ci+d) \quad \text{and} \quad di-b = \lambda' i(-ci+a)$$

with $\lambda,\lambda' > 1$. Therefore

$$a = \lambda d = \lambda\lambda' a \quad \text{and} \quad d = \frac{a}{\lambda} = \frac{d}{\lambda\lambda'} ,$$

whence the first two equations are equivalent with $a = d = 0$ and $|b| > 1$. Thus by (3.16) and (3.17) Lemma 1 (i) holds in this case with $\nu = |b| > 1$.

If $(\xi,\chi) = (\vartheta,\zeta')$ then (3.18) and (3.19) mean that

$$ai + b = \lambda i(ci+d) \ , \ \lambda > 0 \ , \text{ and } \ -\frac{d}{c} = \infty \ .$$

These two equations are equivalent with $b = c = 0$. Therefore Lemma 1(i) holds in this case with $\nu = |d| > 0$.

If $(\xi,\chi) = (\vartheta,\vartheta')$ then (3.18) and (3.19) mean that $\frac{a}{c} = 0$ and $-\frac{d}{c} = 0$, whence $a = d = 0$ and Lemma 1 (i) follows in this case with $\nu = |c| > 0$.

If $(\xi,\chi) = (\eta,\zeta')$ then (3.18) and (3.19) mean that

$$\frac{a^2+b^2}{c^2+d^2} = 1 \quad \text{and} \quad -\frac{d}{c} - \frac{b}{c} = 0\ ,\ |\frac{b}{a}| \geq 1\ ,\ -\frac{b}{a} \neq 1\ .$$

Therefore

$$(\frac{a}{c})^2 = (\frac{a}{c})^2 \frac{1+(b/a)^2}{1+(d/c)^2} = \frac{a^2+b^2}{c^2+d^2} = 1\ ,\ (\frac{b}{d})^2 = (\frac{b}{d})^2 \frac{1+(a/b)^2}{1+(c/d)^2} = 1\ .$$

Since N has determinant 1 it follows that $2ad = 1$ and either $c = a$, $b = -d$ or $c = -a$, $b = d$. If, in addition, $|\frac{b}{a}| > 1$ then $N(i)$ is not on $\mathfrak{F}_\eta$ whence N and a fortiori M belong to ${}_\eta G_{\zeta'}$ by (3.10) and (3.12). Simple matrix multiplications show that Lemma 1(i) then holds with $\nu = \sqrt{2}|d| > 1$ and $\delta = \frac{1+a/c}{2}$. On the other hand, $|a| = |c|$, $|b| = |d|$ and $\frac{b}{a} = 1$ can hold with real entries only if $a = b = -c = d = 1/\sqrt{2}$. This means that $N = R_\eta$. By (3.5) and (3.10) M then belongs to ${}_\eta \mathcal{S}_{\zeta'} = {}_\eta g_{\zeta'}$. Thus Lemma 1 (ii) follows from (3.15) by inversion.

If $(\xi,\chi) = (\eta,\vartheta')$ then (3.18) and (3.19) mean that $|\frac{a}{c}| = 1$ and $-\frac{d}{c} - \frac{b}{a} = 0$, whence $2ad = 1$ and either $c = a$, $b = -d$ or $c = -a$, $b = d$. Similarly as for $(\xi,\chi) = (\eta,\zeta')$ one checks that Lemma 1 (i) holds for $(\xi,\chi) = (\eta,\vartheta')$ with $\nu = \sqrt{2}|a| > 0$ and $\delta = \frac{1-a/c}{2}$.

If $(\xi,\chi) = (\eta,\eta')$ then (3.18) and (3.19) mean that

$$|\frac{ab}{cd}| = 1 \quad \text{and} \quad |\frac{db}{ca}| = 1\ ,$$

whence

$$1 = |\frac{ab}{cd}|\,|\frac{db}{ca}| = (\frac{b}{c})^2 \quad \text{and} \quad 1 = |\frac{ab}{cd}|\,|\frac{ca}{db}| = (\frac{a}{d})^2\ .$$

Since N has determinant 1 this leaves us with the following three possibilities:

$$a = d \ , \quad b = -c \quad \text{with} \quad b^2 + d^2 = 1 \quad \text{or}$$

$$a = d \ , \quad b = c \quad \text{with} \quad -b^2 + d^2 = 1 \quad \text{or}$$

$$a = -d \ , \quad b = -c \quad \text{with} \quad b^2 - d^2 = 1 \ .$$

Now M and a fortiori N are not in $\mathfrak{z}_{\eta\eta'}$, whence none of their entries vanishes by our table for ${}_\xi\mathfrak{z}_\chi$. Thus in the first combination above we must have $N = \pm\exp(\nu X_\zeta)$ with $0 < |\nu - \pi| < \pi$. By (3.5), (3.10) and (3.12) this shows that N and a fortiori M then belong to ${}_\eta g_{\eta'}$. On the other hand, in the second and third combination above $N(0)$ and $N(\infty)$ are non-zero real numbers of equal sign. This is possible only if N and a fortiori M belong to ${}_\eta G_{\eta'}$. Thus we have established Lemma 1 (iii) and now the last entry in our table for ${}_\xi\mathfrak{f}_\chi$ is easily justified. If the second or third combination occurs we set $\delta = \frac{1}{2}(1 - \frac{bd}{|bd|})$ and $\delta' = \frac{1}{2}(1 + \frac{ab}{|bd|})$. Then we obtain from (3.16) by straightforward matrix multiplications

$$R_\vartheta^{-\delta} N R_\vartheta^{\delta'} = \pm \begin{pmatrix} -d' & b' \\ -b' & d' \end{pmatrix}$$

with $b' = \max(|b|,|d|)$, $d' = \min(|b|,|d|)$ and

$$R_\eta^{-1} \begin{pmatrix} \nu & 0 \\ 0 & 1/\nu \end{pmatrix} R_{\eta'} R_\vartheta = \frac{1}{2} \begin{pmatrix} -\nu+1/\nu & \nu+1/\nu \\ -\nu-1/\nu & \nu-1/\nu \end{pmatrix} .$$

Thus Lemma 1 (i) holds for $(\xi,\chi) = (\eta,\eta')$ with $\nu = |b|+|d| > 1$.

Finally for $(\xi,\chi) = (\zeta,\vartheta')$, (ζ,η') or (ϑ,η') Lemma 1 (i) follows from its validity for $(\xi,\chi) = (\vartheta,\zeta')$, (η,ζ') or (η,ϑ') and first part of the proof.

<u>Remark</u>. From an analytical point of view ${}_\xi\Lambda^{\ell}_\chi$, ${}_\xi\Lambda^{r}_\chi$ and ${}_\xi\nu_\chi$ are by no means the only parameters which reveal how G splits up into double cosets $H_\xi M H_\chi$. In particular special definitions as in (3.14) could be avoided. From a geometrical point of view, however, i.e. in terms of points, geodesics and horocycles in $\mathfrak{H}$ the parameters we chose come up very naturally. Indeed they are unique in the sense that

certain families of curves intersect orthogonally and not just transversally. These curves arise by looking at the images of

$$\begin{cases} \{i\}\ , & \text{if } \chi = \zeta'\ , \\ \{z \mid \operatorname{Im} z = 1\}\ , & \text{if } \chi = \vartheta'\ , \\ \{z \mid \operatorname{Re} z = 0\}\ , & \text{if } \chi = \eta'\ , \end{cases}$$

under M in ${}_{\xi}G_{\chi}$ when either ${}_{\xi}\Lambda_{\chi}^{\ell}(M)$ or ${}_{\xi}\nu_{\chi}(M)$ is kept fixed.

In order to construct eigenfunctions of Δ with singularities in $\mathfrak{H}$ we are led to work in $SL_2(\mathbb{C})$. Our next task is to establish some decompositions for this group.

Let

$$\mathfrak{s}_{\chi} = \{M \text{ in } SL_2(\mathbb{C}) \mid M(z) = \infty \text{ for some } z \text{ in } \mathfrak{f}_{\chi}\}. \tag{3.20}$$

By (2.1) and (3.4) $\mathfrak{s}_{\chi}$ consists of those M whose entries satisfy the equations in the following table

χ	ζ'	ϑ'	η'
$\mathfrak{s}_{\chi}$	$c^2+d^2=0$	$c=0$	$cd=0$

For M in $SL_2(\mathbb{C}) - \mathfrak{s}_{\chi}$ we set

$$\Lambda_{\chi}^{\ell}(M) = \frac{1}{\#\mathfrak{f}_{\chi}} \sum_{z \in \mathfrak{f}_{\chi}} z_M\ , \qquad \Lambda_{\chi}^{r}(M) = -w_{\chi}(M^{-1}(\infty))\ , \tag{3.21}$$

and prove

<u>Lemma 2</u>. For every M in $SL_2(\mathbb{C}) - \mathfrak{s}_{\chi}$ there is a unique non-zero complex number $\mu = \mu_{\chi}(M)$ such that

$$M = \exp(\Lambda_{\chi}^{\ell}(M) X_{\vartheta}) \begin{pmatrix} 1/\mu & 0 \\ 0 & \mu \end{pmatrix} R_{\chi} \exp(\Lambda_{\chi}^{r}(M) X_{\chi})\ .$$

<u>Proof</u>: We deduce from (3.4), (3.6) and (3.8) that

$$\Lambda_{\chi}^{\ell}(\exp(\varrho X_{\vartheta}) M \exp(\tau X_{\chi})) = \frac{1}{\#\mathfrak{f}_{\chi}} \sum_{z \in \mathfrak{f}_{\chi}} (z_M + \varrho) = \Lambda_{\chi}^{\ell}(M) + \varrho$$

and

$$\Lambda_\chi^r(\exp(\varrho X_\vartheta)\, M \exp(\tau X_\chi)) = -w_\chi(\exp(-\tau X_\chi) M^{-1}(\infty)) \equiv \Lambda_\chi^r(M) + \tau \pmod{\ell_\chi}$$

for ϱ, τ in $\mathbb{C}$. Therefore

$$N = \exp(-\Lambda_\chi^\ell(M) X_\vartheta)\, M \exp(-\Lambda_\chi^r(M) X_\chi)$$

satisfies

$$\Lambda_\chi^\ell(N) = 0 \quad \text{and} \quad \Lambda_\chi^r(N) \equiv 0 \pmod{\ell_\chi} . \tag{3.22}$$

We express (3.22) by the entries of N which we denote again by a,b,c,d as in the proof of Lemma 1. Then it follows from (3.4), (3.6) and (3.21) that (3.22) holds precisely if

$$\frac{ai+b}{ci+d} + \frac{-ai+b}{-ci+d} = 0 \quad \text{and} \quad -\frac{d}{c} = \infty \text{ , i.e. } \quad b = c = 0 \quad \text{for} \quad \chi = \zeta' ,$$

$$\frac{a}{c} = 0 \quad \text{and} \quad -\frac{d}{c} = 0 \text{ , i.e. } \quad a = d = 0 \quad \text{for} \quad \chi = \vartheta' ,$$

$$\frac{a}{c} + \frac{b}{d} = 0 \quad \text{and} \quad -\frac{d}{c} = 1 \text{ , i.e. } \quad a = b \quad \text{and} \quad c = -d \quad \text{for} \quad \chi = \eta' .$$

Thus Lemma 2 follows from (3.16).

4. Integral representations of eigenfunctions

Let T denote the involution

$$Tf(z) = f(-\bar{z})$$

on functions f defined on $\mathfrak{H}$. We make use of the following symmetry operators

$$S_\xi^{(1)} f(z) = \begin{cases} f(z) , & \text{if} \quad \xi = \zeta, \vartheta \\ f(z) + \varepsilon\, Tf(z) , & \text{if} \quad \xi = (\eta_1, \eta_2, \varepsilon) , \end{cases} \tag{4.1}$$

and

$$S_\xi^{(2)} f(z) = \begin{cases} \varepsilon\, Tf(z) , & \text{if} \quad \xi = (\eta_1, \eta_2, \varepsilon) \quad \text{and} \quad \operatorname{Re} z < 0 , \\ f(z) , & \text{otherwise.} \end{cases} \tag{4.2}$$

We immediately verify that

$$S_\eta^{(\ell)} T = T S_\eta^{(\ell)} = \varepsilon S_\eta^{(\ell)} \tag{4.3}$$

for $\eta = (\eta_1, \eta_2, \varepsilon)$ and $\ell = 1,2$.

In the notation of (2.1) we set

$$Y(M,z) = \frac{y}{(cz+d)(c\bar{z}+d)}$$

for M in $SL_2(\mathbb{C})$ and z in $\mathfrak{H}$. If N is in G then $Y(N,z) = \operatorname{Im} z_N$, whence

$$Y(MN,z) = Y(M,z_N) \tag{4.4}$$

for N in G . Moreover we note that by (3.2)

$$Y(\exp(\tau X_\vartheta)M,z) = Y(M,z) \quad \text{and} \quad Y(\exp(\tau X_\eta)M,z) = e^\tau Y(M,z) \tag{4.5}$$

for τ in $\mathbb{C}$. If we define

$$Y_\xi : (\varrho,z) \longmapsto Y(R_\xi \exp(\varrho X_\xi),z)$$

with R_ξ given by (3.16) it follows from (3.2) and (3.6) that

$$Y_\xi(\varrho,z) = \begin{cases} \dfrac{\operatorname{sh} v_\xi}{2\sin\frac{\varrho+w_\xi}{2}\sin\frac{\varrho+\bar{w}_\xi}{2}}, & \text{if } \xi = \zeta, \\[2ex] \dfrac{v_\xi}{(\varrho+w_\xi)(\varrho+\bar{w}_\xi)}, & \text{if } \xi = \vartheta, \\[2ex] \dfrac{\sin v_\xi}{2\operatorname{sh}\frac{\varrho+w_\xi}{2}\operatorname{sh}\frac{\varrho+\bar{w}_\xi}{2}}, & \text{if } \xi = \eta, \end{cases} \tag{4.6}$$

where $w_\xi = u_\xi + iv_\xi$ stands for $w_\xi(z) = u_\xi(z) + iv_\xi(z)$. Thus $Y_\xi(\varrho,z)$ is analytic in the ϱ-plane except for simple poles at

$$\varrho \equiv -w_\xi \pmod{\ell_\xi} \quad \text{or} \quad \varrho \equiv -\bar{w}_\xi \pmod{\ell_\xi} .$$

Now let $\mathcal{P}_\xi$ denote the path on the real line connecting

$$\begin{cases} 0 \text{ with } 2\pi, & \text{if } \xi = \zeta, \\ -\infty \text{ with } \infty, & \text{if } \xi = \vartheta,\eta, \end{cases}$$

and let $\mathfrak{D}_\xi(z)$ denote the path in the ϱ-plane which connects $-w_\xi$ with $-\bar{w}_\xi$ in a straight line. Note that $Y_\xi(\varrho,z)$ is positive for real ϱ or ϱ on $\mathfrak{D}_\xi(z)$ minus the endpoints. Thus by our convention on logarithms and with $e(\varrho) = e^{2\pi i \varrho}$

$$U_\xi(z,s,\lambda) = S_\xi^{(1)} \int_{\mathcal{C}_\xi} e(-\lambda\varrho)\, Y_\xi^s(\varrho,z)\, d\varrho \tag{4.7}$$

and

$$V_\xi(z,s,\lambda) = S_\xi^{(2)} \frac{1}{i} \int_{\mathfrak{D}_\xi(z)} e(-\lambda\varrho)\, Y_\xi^{1-s}(\varrho,z)\, d\varrho \tag{4.8}$$

are uniquely defined. We always assume that $2\pi\lambda$ is in $\mathbb{Z}$ if $\xi = \zeta$ and that λ is in $\mathbb{R}$ if $\xi = \vartheta, \eta$. Then by (3.6), (4.1) and (4.6) the integral in (4.7) converges absolutely and locally uniformly for z in $\mathfrak{H}$ and $s = \sigma + it$ in

$$\begin{cases} \mathbb{C}\ , & \text{if } \xi = \zeta\ , \\ \text{the half-plane } \sigma > 1/2\ , & \text{if } \xi = \vartheta\ , \\ \text{the half-plane } \sigma > 0\ , & \text{if } \xi = \eta\ . \end{cases}$$

On the other hand, by (3.5), (3.6), (4.2) and (4.6) V_ξ is given by an integral converging absolutely and locally uniformly for s in the half-plane $\sigma > 0$ and z in $\mathfrak{H} - \mathfrak{F}_\xi$.

Since by (2.2)

$$\Delta y^s = s(s-1)\, y^s$$

the G-invariance of Δ yields

$$\Delta Y_\xi^s(\varrho,z) = s(s-1)\, Y_\xi^s(\varrho,z)$$

for ϱ in $\mathbb{R}$. By analytic continuation the above equation continues to hold for ϱ on $\mathfrak{D}_\xi(z)$ minus the endpoints. Thus we obtain

$$\Delta U_\xi(z,s,\lambda) = s(s-1)\, U_\xi(z,s,\lambda)\ , \quad z \text{ in } \mathfrak{H}\ , \tag{4.9}$$

for s in $\sigma > 1/2$ and

$$\Delta V_\xi(z,s,\lambda) = s(s-1)\, V_\xi(z,s,\lambda)\ , \quad z \text{ in } \mathfrak{H} - \mathfrak{F}_\xi\ , \tag{4.10}$$

for s in $\sigma > 0$. If τ is real we infer from (4.4) and the definitions above that

$$U_\xi(\exp(\tau X_\xi)(z),s,\lambda) = S_\xi^{(1)} \int_{\mathcal{b}_\xi} e(-\lambda\varrho) Y_\xi^s(\varrho+\tau,z)\,d\varrho = e(\lambda\tau) U_\xi(z,s,\lambda) \quad (4.11)$$

and that

$$V_\xi(\exp(\tau X_\xi)(z),s,\lambda) = S_\xi^{(2)} \frac{1}{i} \int_{\mathfrak{D}_\xi(z)-\tau} e(-\lambda\varrho) Y_\xi^{1-s}(\varrho+\tau,z)\,d\varrho = e(\lambda\tau) V_\xi(z,s,\lambda) \quad (4.12)$$

since $\mathfrak{D}_\xi(\exp(\tau X_\xi)(z)) \equiv \mathfrak{D}_\xi(z)-\tau$ (mod Re ℓ_ξ) by (3.8). Finally we note that for $\eta = (\eta_1,\eta_2,\varepsilon)$

$$TU_\eta(z,s,\lambda) = \varepsilon U_\eta(z,s,\lambda) \quad \text{and} \quad TV_\eta(z,s,\lambda) = \varepsilon V_\eta(z,s,\lambda) \quad (4.13)$$

by (4.3) while

$$U_\eta(-\frac{1}{z},s,\lambda) = \varepsilon U_\eta(z,s,-\lambda) \quad \text{and} \quad V_\eta(-\frac{1}{z},s,\lambda) = \varepsilon V_\eta(z,s,-\lambda) \quad (4.14)$$

since

$$Y_\eta(\varrho,-1/z) = Y_\eta(-\varrho,-\bar{z})$$

by (3.6) and (4.6).

We shall show that (4.9)-(4.13) essentially characterize the functions U_ξ and V_ξ. First, however, we determine their boundary behaviour. In doing this we make constant use of (3.6) and (4.6)-(4.8).

We start with the simplest cases, namely

$$U_\vartheta(z,s,0) = \int_{-\infty}^{\infty} \left(\frac{y}{|z+\varrho|^2}\right)^s d\varrho = y^{1-s}\int_{-\infty}^{\infty}(1+\varrho^2)^{-s} d\varrho = v_\vartheta^{1-s} B(1/2,s-1/2), \quad (4.15)$$

if $\sigma > 1/2$, and

$$V_\vartheta(z,s,0) = y^s \int_{-1}^{1} (1-\varrho^2)^{s-1} d\varrho = v_\vartheta^s B(1/2,s), \quad (4.16)$$

if $\sigma > 0$, where

$$B(\alpha,\beta) = \frac{\Gamma(\alpha)\Gamma(\beta)}{\Gamma(\alpha+\beta)} = \int_0^1 \varrho^{\alpha-1}(1-\varrho)^{\beta-1} d\varrho$$

and $\Gamma(s)$ denotes the gamma-function. For arbitrary λ we have

$$U_\vartheta(z,s,\lambda) = e(\lambda x) y^{1-s} \int_{-\infty}^{\infty} e(-\lambda y\tau)(1+\tau^2)^{-s} d\tau \sim e(\lambda u_\vartheta) v_\vartheta^{1-s} B(1/2,s-1/2), \quad (4.17)$$

if $\sigma > 1/2$ and $v_\vartheta \to 0$, and

$$U_\zeta(z,s,\lambda) = e(\lambda u_\zeta) \int_0^{2\pi} e(-\lambda\varrho)\left(\frac{\operatorname{sh} v_\zeta}{2\sin\frac{\varrho+w_\zeta}{2}\sin\frac{\varrho+\bar{w}_\zeta}{2}}\right)^s d\varrho \sim e(\lambda u_\zeta)(2v_\zeta)^s \tag{4.18}$$

$$\int_0^{2\pi} e(-|\lambda|\varrho)|1-e^{i\varrho}|^{-2s} d\varrho = e(\lambda u_\zeta) v_\zeta^s\, 2^{s+1} \sin\pi s \;\; B(s+2\pi|\lambda|, 1-2s) ,$$

if $\sigma < 1/2$ and $v_\zeta \to 0$. For the last integral in (4.18) equals

$$\frac{1}{i} \int_{|\tau|=1} \tau^{2\pi|\lambda|-1}((1-\tau)(1-1/\tau))^{-s} d\tau ,$$

where the unit circle is run through anticlockwise. Since by our assumptions $2\pi|\lambda|$ is a non-negative integer for $\xi = \zeta$ we may shift the integration from the unit circle to a path winding closely around the interval $[0,1]$. Taking into account that $\log \frac{(1-\tau)^2}{-\tau}$ was real on $|\tau| = 1$ we can express the above integral by

$$\frac{1}{i}\int_0^1 \tau^{2\pi|\lambda|+s-1}(1-\tau)^{-2s} d\tau\,(e(\tfrac{s}{2})-e(-\tfrac{s}{2})) = 2\sin\pi s \;B(2\pi|\lambda|+s, 1-2s)$$

for $0<\sigma<1/2$, whence (4.18) follows. On the other hand,

$$(v_\xi(1-\tau^2))^{1-s} Y_\xi^{1-s}(-u_\xi+iv_\xi\tau, z) = \beta_\xi^{1-s} + O(v_\xi) ,\; v_\xi \to \infty ,$$

uniformly for $-1 \leqslant \tau \leqslant 1$, where

$$\beta_\xi = \begin{cases} 1 & , \text{ if } \xi = \vartheta , \\ 2 & , \text{ if } \xi = \zeta \text{ or } \eta . \end{cases} \tag{4.19}$$

Thus we obtain for $\sigma > 0$

$$\begin{aligned} V_\xi(z,s,\lambda) &= e(\lambda u_\xi) v_\xi^s \int_{-1}^1 (1-\tau^2)^{s-1}(\beta_\xi^{1-s} + O(v_\xi))\, d\tau \\ &= e(\lambda u_\xi) v_\xi^s B(1/2, s)\, \beta_\xi^{1-s} + O(v_\xi^{\sigma+1}) ,\; v_\xi \to 0 . \end{aligned} \tag{4.20}$$

It is straightforward that

$$U_\zeta(z,s,0) = (1-e^{-2v_\zeta})^s \int_0^{2\pi} |1-e^{i\varrho-v_\zeta}|^{-2s} d\varrho = 2\pi + O(e^{-v_\zeta}) ,\; v_\zeta \to \infty , \tag{4.21}$$

while for $\sigma > 0$

$$V_\zeta(z,s,0)=2\left(1-e^{-2v_\zeta}\right)^{1-s}\int_0^{v_\zeta}\left[\left(1-e^{-\tau-v_\zeta}\right)\left(1-e^{\tau-v_\zeta}\right)\right]^{s-1}d\tau =$$

$$= 2\int_0^{v_\zeta}(1-e^{-\tau})^{s-1}d\tau\left(1+O\left(e^{-v_\zeta}\right)\right)$$

$$= 2v_\zeta\left(1-e^{-v_\zeta}\right)^{s-1}-2(s-1)\int_0^\infty \tau e^{-\tau}(1-e^{-\tau})^{s-2}d\tau + O\left(v_\zeta e^{-v_\zeta}\right) \qquad (4.22)$$

$$= 2v_\zeta - 2(s-1)\frac{\partial}{\partial z}B(z,s-1)\Big|_{z=1} + O\left(v_\zeta e^{-v_\zeta}\right) =$$

$$= 2\left(v_\zeta + \Gamma'(1) - \frac{\Gamma'}{\Gamma}(s)\right) + O\left(v_\zeta e^{-v_\zeta}\right), v_\zeta \to \infty .$$

Now let $\lambda \neq 0$ and $\mathcal{B}_\xi(z,\pm 1)$ denote the path in the ϱ-plane that connects $-u_\xi \pm iv_\xi$ with $-u_\xi \pm i\infty$ in a straight line. By Cauchy's theorem, (4.6) and the periodicity of the integrand for $\xi = \zeta$ we may shift the integration in (4.7) from $\mathcal{B}_\xi$ to a loop winding around $\mathcal{B}_\xi(z,-\frac{\lambda}{|\lambda|})$. By shrinking this loop to $\mathcal{B}_\xi(z,-\frac{\lambda}{|\lambda|})$ we can express U_ξ by an integral over $\mathcal{B}_\xi(z,-\frac{\lambda}{|\lambda|})$. Recalling that $\log Y_\xi(\varrho,z)$ was real on $\mathcal{B}_\xi$ we thus obtain

$$U_\zeta(z,s,\lambda) = \frac{e(\lambda u_\zeta)}{i}\left(e\left(\frac{s}{2}\right)-e\left(-\frac{s}{2}\right)\right)\int_{v_\zeta}^\infty e^{-2\pi|\lambda|\tau}\left(\frac{1-e^{-2v_\zeta}}{(1-e^{-\tau-v_\zeta})(e^{\tau-v_\zeta}-1)}\right)^s d\tau$$

and (4.23)

$$U_\vartheta(z,s,\lambda) = \frac{e(\lambda u_\vartheta)}{i}\left(e\left(\frac{s}{2}\right)-e\left(-\frac{s}{2}\right)\right)\int_{v_\vartheta}^\infty e^{-2\pi|\lambda|\tau}\left(\frac{v_\vartheta}{(\tau+v_\vartheta)(\tau-v_\vartheta)}\right)^s d\tau .$$

These integrals converge absolutely and locally uniformly for s in $-1<\sigma<1$ since $2\pi|\lambda| \geqslant 1$ if $\xi = \zeta$ and $\lambda \neq 0$. Moreover the first of the above integrals is

$$\sim e^{-2\pi|\lambda|v_\zeta}\int_0^\infty e^{-2\pi|\lambda|\tau}(e^\tau-1)^{-s}d\tau = e^{-2\pi|\lambda|v_\zeta}\int_0^1 \varrho^{2\pi|\lambda|-1}(\tfrac{1}{\varrho}-1)^{-s}d\varrho$$

$$= e^{-2\pi|\lambda|v_\zeta} B(2\pi|\lambda|+s,1-s) \quad , \quad v_\zeta \to \infty \; ,$$

while the second is

$$\sim e^{-2\pi|\lambda|v_\vartheta}\int_0^\infty e^{-2\pi|\lambda|\tau}(2\tau)^{-s}d\tau = e^{-2\pi|\lambda|v_\vartheta}\Gamma(1-s)\,\frac{(\pi|\lambda|)^{s-1}}{2} \quad , \quad v_\vartheta \to \infty \; .$$

By well-known functional equations for $\Gamma(s)$ we therefore have

$$U_\xi(z,s,\lambda) \sim e(\lambda u_\xi)\,e^{-2\pi|\lambda|v_\xi}\begin{cases}(|\lambda|B(2\pi|\lambda|,s))^{-1} & \text{for } \xi=\zeta \; , \\[2ex] \dfrac{\pi}{\Gamma(s)}(\pi|\lambda|)^{s-1} & \text{for } \xi=\vartheta \; ,\end{cases} \tag{4.24}$$

of $\lambda \neq 0$, $-1<\sigma<1$ and $v_\xi \to \infty$. On the other hand, if $\lambda \neq 0$, $\sigma > 0$ and $v_\xi \to \infty$

$$\begin{aligned} V_\xi(z,s,\lambda) &= e(\lambda u_\xi)\int_{-v_\xi}^{v_\xi} e^{2\pi\lambda\tau}\, Y_\xi^{1-s}(-u_\xi+i\tau,z)\,d\tau \\ &\sim e(\lambda u_\xi)\,e^{2\pi|\lambda|v_\xi}\int_0^{v_\xi} e^{-2\pi|\lambda|\tau}\begin{Bmatrix}(1-e^{-\tau})^{s-1} & \text{for } \xi=\zeta \\ (2\tau)^{s-1} & \text{for } \xi=\vartheta\end{Bmatrix} d\tau \\ &\sim e(\lambda u_\xi)\,e^{2\pi|\lambda|v_\xi}\begin{cases}B(2\pi|\lambda|,s) & \text{for } \xi=\zeta \; , \\[1ex] \dfrac{\Gamma(s)}{2}(\pi|\lambda|)^{-s} & \text{for } \xi=\vartheta \; .\end{cases}\end{aligned} \tag{4.25}$$

The last line follows by extending the integration in the preceding integral from $(0,v_\xi)$ to $(0,\infty)$.

Finally we consider the behaviour of U_η and V_η as $v_\eta \to \frac{\pi}{2}$. Noting that

$$Y_\eta^s(\varrho,z)\,|_{v_\eta=\pi/2} = Y_\eta^s(\varrho,iy) = [\mathrm{ch}(\varrho+u_\eta)]^{-s}$$

and

$$\frac{\partial}{\partial v_\eta} Y_\eta^s(\varrho,z)\,|_{v_\eta=\pi/2} = -s[\mathrm{ch}(\varrho+u_\eta)]^{-s-1} = -s\; Y_\eta^{s+1}(\varrho,iy)$$

we have for $\sigma>0$

$$\int_{\mathcal{C}_\eta} e(-\lambda\varrho)Y_\eta^s(\varrho,iy)\,d\varrho = e(\lambda u_\eta)\int_{-\infty}^{\infty} e(-\lambda\varrho)\,(\mathrm{ch}\varrho)^{-s}d\varrho =$$

$$= e(\lambda u_\eta)\,2^{s-1}B(\tfrac{s}{2}+\pi i\lambda,\ \tfrac{s}{2}-\pi i\lambda)$$

by the top formula on p. 10 in [14] and

$$\frac{1}{i}\int_{\mathfrak{D}_\eta(iy)} e(-\lambda\varrho)Y_\eta^{1-s}(\varrho,iy)\,d\varrho = e(\lambda u_\eta)\int_{-\pi/2}^{\pi/2} e^{2\pi\lambda\tau}(\cos\tau)^{s-1}d\tau$$

$$= e(\lambda u_\eta)\,\frac{\pi 2^{1-s}}{sB(\frac{1+s}{2}+\pi i\lambda,\frac{1+s}{2}-\pi i\lambda)}$$

by the top formula on p. 9 in [14]. Incidentally these integrals can also be evaluated similarly as the integral in (4.18). For $\eta=(\eta_1,\eta_2,\varepsilon)$ and $\sigma>0$ Taylor's formula and (4.1) now yield

$$U_\eta(z,s,\lambda) = \int_{\mathcal{C}_\eta} e(-\lambda\varrho)\,[\,(1+\varepsilon)Y_\eta^s(\varrho,iy) + (1-\varepsilon)s\;Y_\eta^{s+1}(\varrho,iy)(\tfrac{\pi}{2}-v_\eta)\,]\,d\varrho + O(|\tfrac{\pi}{2}-v_\eta|^2)$$

$$= e(\lambda u_\eta)(1+\varepsilon)2^{s+1}B(\tfrac{s}{2}+i\pi\lambda,\tfrac{s}{2}-i\pi\lambda)$$

$$+\ e(\lambda u_\eta)(\tfrac{\pi}{2}-v_\eta)(1-\varepsilon)2^s s\;B(\tfrac{1+s}{2}+\pi i\lambda,\tfrac{1+s}{2}-\pi i\lambda)$$

$$+\ O(|\tfrac{\pi}{2}-v_\eta|^2),\quad v_\eta\to\tfrac{\pi}{2}, \tag{4.26}$$

while (4.2) similarly leads to

$$V_\eta(z,s,\lambda) = \left(\frac{\frac{\pi}{2}-v_\eta}{|\frac{\pi}{2}-v_\eta|}\right)^{\frac{1+\varepsilon}{2}} \frac{1}{i}\int_{\mathfrak{D}_\eta(iy)} e(-\lambda\varrho)\,[Y_\eta^{1-s}(\varrho,iy) + (1-s)Y_\eta^{2-s}(\varrho,iy)\,|\tfrac{\pi}{2}-v_\eta|\,]\,d\varrho$$

$$+\ O(|\tfrac{\pi}{2}-v_\eta|^2) = \left(\frac{\frac{\pi}{2}-v_\eta}{|\frac{\pi}{2}-v_\eta|}\right)^{\frac{1+\varepsilon}{2}} e(\lambda u_\eta)\left[\frac{\pi\,2^{1-s}}{sB(\frac{1+s}{2}+i\pi\lambda,\frac{1+s}{2}-i\pi\lambda)} - \frac{\pi 2^{2-s}|\frac{\pi}{2}-v_\eta|}{B(\frac{s}{2}+i\pi\lambda,\frac{s}{2}-i\pi\lambda)}\right]$$

$$+\ O(|\tfrac{\pi}{2}-v_\eta|^2),\quad v_\eta\to\tfrac{\pi}{2}. \tag{4.27}$$

Our next lemma shows in what sense (4.9) - (4.13) characterize U_ξ and V_ξ.

Lemma 3. Let f be a C^∞-function on $\mathfrak{H}-\mathfrak{F}_\xi$, where $\mathfrak{F}_\xi$ is given by (3.5). Let $2\pi\lambda$ be in $\mathbb{Z}$ for $\xi = \zeta$ and in $\mathbb{R}$ for $\xi = \vartheta$ or η. Suppose that

(i) $\quad f(\exp(\tau X_\xi)(z)) = e(\lambda\tau) f(z)$, τ real,

and for a complex number s with $\frac{1}{2} \leq \sigma < 1$

(ii) $\quad \Delta f(z) + s(1-s) f(z) = 0$, z in $\mathfrak{H} - \mathfrak{F}_\xi$.

If $\xi = (\eta_1, \eta_2, \varepsilon)$, assume moreover that

(iii) $Tf = \varepsilon f$.

Then there are complex numbers c_1, c_2 such that

$$f(z) = c_1 U_\xi(z,s,\lambda) + c_2 V_\xi(z,s,\lambda), \quad z \text{ in } \mathfrak{H} - \mathfrak{F}_\xi$$

unless $\xi = \vartheta$, $\lambda = 0$ and $s = 1/2$.

Proof. Set $g(u_\xi, v_\xi) = f(z)$ where $u_\xi + iv_\xi = w_\xi(z)$. By (i) and (3.8) we have

$$g(\tau + u_\xi, v_\xi) = e(\lambda\tau) g(u_\xi, v_\xi),$$

whence

$$f(z) = e(\lambda u_\xi) g(0, v_\xi).$$

Now (ii) and (3.6) imply that the function

$$v_\xi \mapsto g(0, v_\xi)$$

satisfies an ordinary linear differential equation of second order. Therefore every function satisfying (i)-(iii) must be a linear combination of any two linearly independent functions with these three properties. By (4.7), (4.8) and (4.23) $U_\xi(z,s,\lambda)$ and $V_\xi(z,s,\lambda)$ are analytic functions of s in $\sigma > 0$ except for $U_\vartheta(z,s,0)$. The latter function is analytic in $\sigma > 0$ except for a simple pole at $s = 1/2$ by (4.15). By analytic continuation (4.9) - (4.13) show that

$U_\xi(z,s,\lambda)$ and $V_\xi(z,s,\lambda)$ fulfill (i)-(iii) in $\sigma>0$ unless $\xi=\vartheta$, $\lambda=0$ and $s=1/2$. With the exception of the case just mentioned they are also linearly independent in view of (4.15), (4.16), (4.21), (4.22) and (4.24) - (4.27). Thus the lemma follows.

Remark: Lemma 3 is meaningless for $\xi=\vartheta$, $\lambda=0$ and $s=1/2$ because $U_\vartheta(z,s,0)$ has a pole at $s=1/2$. However it can be upheld if in the statement $U_\vartheta(z,1/2,0)$ is replaced by $y^{1/2}\log y$ since $V_\vartheta(z,1/2,0)=\pi y^{1/2}$.

As an application of the above characterization we obtain

Lemma 4. The functions $U_\xi(z,s,\lambda)$ and $V_\xi(z,s,\lambda)$ are meromorphic in s. They satisfy the following functional equations

$$U_\xi(z,1-s,\lambda)=\begin{cases}\operatorname{tg}\pi s\, V_\xi(z,s,\lambda) & \text{, if } \xi=\vartheta \text{ and } \lambda=0,\\ \gamma_\xi(s,\lambda)U_\xi(z,s,\lambda) & \text{, otherwise,}\end{cases}$$

and

$$V_\xi(z,1-s,\lambda) = -\operatorname{ctg}\pi s\, U_\xi(z,s,\lambda)+\gamma_\xi(1-s,\lambda)V_\xi(z,s,\lambda),$$

where

$$\gamma_\xi(s,\lambda)=\begin{cases}\dfrac{\Gamma(s)\Gamma(1-s+2\pi|\lambda|)}{\Gamma(1-s)\Gamma(s+2\pi|\lambda|)} & \text{, if } \xi=\zeta,\\ 0 & \text{, if } \xi=\vartheta \text{ and } \lambda=0\\ \dfrac{\Gamma(s)}{\Gamma(1-s)}(\pi|\lambda|)^{1-2s} & \text{, if } \xi=\vartheta \text{ and } \lambda\neq 0,\\ 2^{1-2s}\dfrac{\Gamma(s)\Gamma(\frac{3-\varepsilon-2s}{4}+i\pi\lambda)\Gamma(\frac{3-\varepsilon-2s}{4}-i\pi\lambda)}{\Gamma(1-s)\Gamma(\frac{1-\varepsilon+2s}{4}+i\pi\lambda)\Gamma(\frac{1-\varepsilon+2s}{4}-i\pi\lambda)} & \text{, if } \xi=(\eta_1,\eta_2,\varepsilon).\end{cases}$$

Proof: In view of (4.9)-(4.13) $U_\xi(z,1-s,\lambda)$ and $V_\xi(z,1-s,\lambda)$ satisfy all assumptions of Lemma 3 if $\frac{1}{2}<\sigma<1$. Thus

$$U_\xi(z,1-s,\lambda) = c_1U_\xi(z,s,\lambda)+c_2V_\xi(z,s,\lambda) \tag{4.28}$$

and

$$V_\xi(z,1-s,\lambda) = c_3U_\xi(z,s,\lambda)+c_4V_\xi(z,s,\lambda) \tag{4.29}$$

for z in $\mathfrak{H}-\mathfrak{F}_\xi$, where the coefficients $c_1,\dots,c_4$ are independent of z. These coefficients depend analytically on s since $U_\xi(z,s,\lambda)$ and $V_\xi(z,s,\lambda)$ are linearly independent. By analytic continuation (4.28) and (4.29) hold for s in the strip $0<\sigma<1$ except possibly for $s = 1/2$. Thus it suffices to show for every ξ that

$$c_1 = \gamma_\xi(s,\lambda) \qquad c_2 = \begin{cases} \mathrm{tg}\pi s & , \text{ if } \xi = \vartheta \text{ and } \lambda = 0, \\ 0 & , \text{ otherwise}, \end{cases}$$

$$c_3 = -\mathrm{ctg}\pi s \qquad c_4 = \gamma_\xi(1-s,\lambda)$$

for s in a non-empty open set in $0<\sigma<1$. We shall accomplish this by comparing the behaviour of both sides in (4.28) and (4.29) as z approaches the boundary of $\mathfrak{H}-\mathfrak{F}_\xi$.

First we have by (4.15) and (4.16)

$$v_\vartheta^s B(1/2,1/2-s) = c_1 v_\vartheta^{1-s} B(1/2,s-1/2) + c_2 v_\vartheta^s B(1/2,s)$$

and

$$v_\vartheta^{1-s} B(1/2,1-s) = c_3 v_\vartheta^{1-s} B(1/2,s-1/2) + c_4 v_\vartheta^s B(1/2,s) ,$$

whence $c_1 = c_4 = 0$ and by the functional equations for $\Gamma(s)$

$$c_2 = \frac{\Gamma(1/2-s)\,\Gamma(1/2+s)}{\Gamma(1-s)\,\Gamma(s)} = \mathrm{tg}\pi s \qquad c_3 = \frac{\Gamma(1-s)\,\Gamma(s)}{\Gamma(3/2-s)\,\Gamma(s-1/2)} = -\mathrm{ctg}\pi s .$$

Thus the lemma holds for $\xi = \vartheta$ and $\lambda = 0$. Since by (4.21), (4.22), (4.28) and (4.29)

$$2\pi \sim c_1 2\pi + c_2 2v_\zeta \qquad \text{and} \qquad 2v_\zeta \sim c_3 2\pi + c_4 2v_\zeta$$

as $v_\zeta \to \infty$ we obtain $c_1 = c_4 = 1$ and $c_2 = 0$. Using this we infer from (4.18)-(4.20) and (4.29) that

$$c_3 2^{s+1} \sin\pi s\, B(s,1-2s) + 2^{1-s} B(1/2,s) = o(1)$$

if $0 < \sigma < 1/2$ and $v_\zeta \to 0$. Hence by the functional equations for $\Gamma(s)$

$$c_3 = -\frac{2^{-2s}\pi^{1/2}\Gamma(1-s)}{\sin\pi s\ \Gamma(s+1/2)\Gamma(1-2s)} = -\mathrm{ctg}\pi s$$

and the lemma is proved for $\xi = \zeta$ and $\lambda = 0$.

If $\xi = \zeta$, $\lambda \neq 0$ and $v_\zeta \to \infty$ (4.24), (4.25), (4.28) and (4.29) yield

$$(|\lambda| B(2\pi|\lambda|,1-s))^{-1} \sim c_1 (|\lambda| B(2\pi|\lambda|,s))^{-1} + c_2 B(2\pi|\lambda|,s) e^{4\pi|\lambda| v_\zeta}$$

and

$$B(2\pi|\lambda|,1-s) \sim c_3 (|\lambda| B(2\pi|\lambda|,s) e^{4\pi|\lambda| v_\zeta})^{-1} + c_4 B(2\pi|\lambda|,s) .$$

Thus $c_2 = 0$,

$$c_1 = \frac{\Gamma(s)\Gamma(1-s+2\pi|\lambda|)}{\Gamma(1-s)\Gamma(s+2\pi|\lambda|)} = \gamma_\zeta(s,\lambda) \quad \text{and} \quad c_4 = \gamma_\zeta(1-s,\lambda) .$$

We determine c_3 from (4.18)-(4.20) and (4.29) which give

$$c_3 2^{s+1} \sin\pi s\, B(2\pi|\lambda|+s,1-2s) + \gamma_\zeta(1-s,\lambda) 2^{1-s} B(1/2,s) = o(1)$$

if $0 < \sigma < 1/2$ and $v_\zeta \to 0$. Again we obtain

$$c_3 = -\frac{2^{-2s}\pi^{1/2}\Gamma(1-s)}{\sin\pi s\ \Gamma(s+1/2)\Gamma(1-2s)} = -\mathrm{ctg}\pi s$$

thereby establishing the lemma for $\xi = \zeta$.

If $\xi = \vartheta$, $\lambda \neq 0$ and $v_\vartheta \to \infty$ (4.24), (4.25), (4.28) and (4.29) imply

$$\frac{\pi}{\Gamma(1-s)}(\pi|\lambda|)^{-s} \sim c_1 \frac{\pi}{\Gamma(s)}(\pi|\lambda|)^{s-1} + c_2 \frac{\Gamma(s)}{2}(\pi|\lambda|)^{-s} e^{4\pi|\lambda| v_\vartheta}$$

and

$$\frac{\Gamma(1-s)}{2}(\pi|\lambda|)^{s-1} \sim c_3 \frac{\pi}{\Gamma(s)}(\pi|\lambda|)^{s-1} e^{-4\pi|\lambda| v_\vartheta} + c_4 \frac{\Gamma(s)}{2}(\pi|\lambda|)^{-s},$$

whence $c_2 = 0$,

$$c_1 = \frac{\Gamma(s)}{\Gamma(1-s)}(\pi|\lambda|)^{1-2s} = \gamma_\vartheta(s,\lambda) \quad \text{and} \quad c_4 = \gamma_\vartheta(1-s,\lambda) .$$

Consequently (4.17), (4.19), (4.20) and (4.29) yield

$$B(1/2,1-s) \sim c_3 B(1/2,s-1/2) + \gamma_\vartheta(1-s,\lambda) B(1/2,s) v_\vartheta^{2s-1} ,$$

if $1/2 < \sigma < 1$ and $v_\vartheta \to 0$. Hence

$$c_3 = \frac{\Gamma(1-s)\,\Gamma(s)}{\Gamma(3/2-s)\,\Gamma(s-1/2)} = -\mathrm{ctg}\,\pi s$$

and the lemma follows for $\xi = \vartheta$.

Finally, for $\xi = (\eta_1, \eta_2, \varepsilon)$ we deduce from (4.26)-(4.29)

$$(1+\varepsilon)2^{-s} B(\tfrac{1-s}{2} + i\pi\lambda, \tfrac{1-s}{2} - i\pi\lambda) + (1-\varepsilon)2^{1-s}(1-s) B(1-\tfrac{s}{2} + i\pi\lambda, 1-\tfrac{s}{2} - i\pi\lambda)(\tfrac{\pi}{2} - v_\eta) + O(|\tfrac{\pi}{2} - v_\eta|^2)$$

$$= c_1[(1+\varepsilon)2^{s-1} B(\tfrac{s}{2} + i\pi\lambda, \tfrac{s}{2} - i\pi\lambda) + (1-\varepsilon)2^s s B(\tfrac{1+s}{2} + i\pi\lambda, \tfrac{1+s}{2} - i\pi\lambda)(\tfrac{\pi}{2} - v_\eta)] +$$

$$+ c_2\left[\frac{\pi 2^{1-s}}{sB(\frac{1+s}{2} + i\pi\lambda, \frac{1+s}{2} - i\pi\lambda)} - \frac{\pi 2^{2-s}(\frac{\pi}{2} - v_\eta)}{B(\frac{s}{2} + i\pi\lambda, \frac{s}{2} - i\pi\lambda)}\right] + O(|\tfrac{\pi}{2} - v_\eta|^2) , \quad v_\eta \nearrow \tfrac{\pi}{2} ,$$

and

$$\frac{\pi 2^s}{(1-s)B(1-\frac{s}{2} + i\pi\lambda, 1-\frac{s}{2} - i\pi\lambda)} - \frac{\pi 2^{1+s}(\frac{\pi}{2} - v_\eta)}{B\left(\frac{1-s}{2} + i\pi\lambda, \frac{1-s}{2} - i\pi\lambda\right)} + O(|\tfrac{\pi}{2} - v_\eta|^2)$$

$$= c_3[(1+\varepsilon)2^{s-1}B(\tfrac{s}{2}+i\pi\lambda,\tfrac{s}{2}-i\pi\lambda) + (1-\varepsilon)2^s sB(\tfrac{1+s}{2}+i\pi\lambda,\tfrac{1+s}{2}-i\pi\lambda)(\tfrac{\pi}{2}-v_\eta)]$$

$$+ c_4\left[\frac{\pi 2^{1-s}}{sB(\frac{1+s}{2}+i\pi\lambda,\frac{1+s}{2}-i\pi\lambda)} - \frac{\pi 2^{2-s}(\frac{\pi}{2}-v_\eta)}{B(\frac{s}{2}+i\pi\lambda,\frac{s}{2}-i\pi\lambda)}\right] + O(|\tfrac{\pi}{2}-v_\eta|^2),\ v_\eta \nearrow \tfrac{\pi}{2}.$$

Since $\varepsilon = \pm 1$ the first of these equations gives $c_2 = 0$ and

$$c_1 = 2^{1-2s}\frac{\Gamma(s)\,\Gamma(\frac{3-\varepsilon-2s}{4}+i\pi\lambda)\,\Gamma(\frac{3-\varepsilon-2s}{4}-i\pi\lambda)}{\Gamma(1-s)\,\Gamma(\frac{1-\varepsilon+2s}{4}+i\pi\lambda)\,\Gamma(\frac{1-\varepsilon+2s}{4}-i\pi\lambda)} = \gamma_\eta(s,\lambda)\ ,$$

while the second yields $c_4 = \gamma_\eta(1-s,\lambda)$. Therefore we obtain

$$c_3 = \frac{\pi}{(1-s)B(1-\frac{s}{2}+i\pi\lambda,1-\frac{s}{2}-i\pi\lambda)B(\frac{s}{2}+i\pi\lambda,\frac{s}{2}-i\pi\lambda)} -$$

$$- \frac{\pi}{sB(\frac{1+s}{2}+i\pi\lambda,\frac{1+s}{2}-i\pi\lambda)B(\frac{1-s}{2}+i\pi\lambda,\frac{1-s}{2}-i\pi\lambda)}$$

$$= \frac{1}{\sin\pi s}[\sin\pi(\tfrac{s}{2}+i\pi\lambda)\sin\pi(\tfrac{s}{2}-i\pi\lambda) - \cos\pi(\tfrac{s}{2}+i\pi\lambda)\cos\pi(\tfrac{s}{2}-i\pi\lambda)] =$$

$$= -\operatorname{ctg}\pi s\ .$$

This completes the proof of Lemma 4.

5. Fourier coefficients and Kloosterman sums

Let Γ and ξ be as in section 2. We denote the stabilizer of ξ in Γ by Γ_ξ . By the convention adopted in section 2 this means in particular that Γ_η consists of those M in Γ which keep $\eta = (\eta_1,\eta_2,\varepsilon)$ fixed componentwise. For each ξ we now select a matrix M_ξ in G such that

$$M_\xi(\xi) = \begin{cases} i\ , & \text{if } \xi = \zeta\ ,\\ \infty\ , & \text{if } \xi = \vartheta\ ,\\ (0,\infty,\varepsilon)\ , & \text{if } \xi = (\eta_1,\eta_2,\varepsilon)\ . \end{cases} \tag{5.1}$$

Then $\Gamma'_\xi = M_\xi\Gamma_\xi M_\xi^{-1}$ is the stabilizer of $M_\xi(\xi)$ in the conjugate group $M_\xi\Gamma M_\xi^{-1}$. Thus there exists a uniquely determined positive number λ_ξ

on $\mathscr{P}_\xi$ such that Γ'_ξ is generated by one or both elements $\pm \exp(\lambda_\xi X_\xi)$. While Γ_ξ determines λ_ξ for $\xi = \zeta, \eta$ we notice that there always exists on M_ϑ yielding $\lambda_\vartheta = 1$. Assuming that M_ϑ has been chosen in this way we have

$$\lambda_\xi = \begin{cases} \dfrac{2\pi}{[\Gamma_\zeta : Z_\Gamma]} , & \text{if } \xi = \zeta , \\ 1 , & \text{if } \xi = \vartheta \\ \log N(P) , & \text{if } \xi = \eta , \end{cases} \tag{5.2}$$

where Z_Γ denotes the center of Γ and $N(P)$ denotes the norm of a primitive P in Γ with $P(\eta) = \eta$ (for the terminology see e.g. [20] p. 73).

The requirements (5.1) and (5.2) do not determine M_ξ uniquely. However, it is easy to see that any other matrix in G satisfying them is necessarily of the form

$$\pm \exp(\varrho X_\xi) M_\xi \tag{5.3}$$

for some ϱ in $\mathbb{R}$. If there is an M in Γ with $\chi = M(\xi)$ then by (5.3)

$$M_\xi = \pm \exp(\varrho X_\xi) M_\chi M$$

for some real ϱ and $\lambda_\xi = \lambda_\chi$. Moreover ϱ is now uniquely determined modulo λ_ξ. For, if also $\chi = N(\xi)$ with N in Γ, then MN^{-1} belongs to Γ_χ. Therefore $M_\chi MN^{-1} M_\chi^{-1} = \pm \exp(n\lambda_\chi X_\chi)$

for some integer n, whence

$$M_\xi = \pm \exp(\varrho X_\xi) M_\chi M = \pm \exp([\varrho+n] X_\xi) M_\chi N .$$

Thus it is compatible with (5.1) and (5.2) if, in addition, we demand that

$$M_\xi^{-1} M_\chi \text{ is in } \Gamma \text{ for } \xi \sim \chi , \tag{5.4}$$

where $\sim$ is the equivalence relation (2.6).

In the next lemma we show that the Γ-invariant eigenfunctions e_j have a Fourier series expansion with respect to every ξ. Condition

(5.4) guarantees that the corresponding Fourier coefficients agree if $\xi \approx \chi$.

Lemma 5. For every ξ and $j \geq 0$ there are complex numbers $\alpha_{j\xi}(n)$ such that

$$S_{\xi}^{(1)} e_j(M_{\xi}^{-1}(z)) = \sum_{n\in Z} \alpha_{j\xi}(n) U_{\xi}(z,s_j, \frac{n}{\lambda_{\xi}}) ,$$

where s_j is related to e_j by (2.9) and the series on the right converges absolutely and locally uniformly on $\mathfrak{H}$. Moreover $\alpha_{j\vartheta}(0) = 0$ if $\operatorname{Re} s_j = 1/2$ and $\alpha_{o\xi}(n) = 0$ if $n \neq 0$, while

$$\alpha_{o\xi}(0) = \begin{cases} (2\pi\omega^{1/2}(\mathfrak{F}))^{-1} & , \text{ if } \xi = \zeta , \\ (\pi\omega^{1/2}(\mathfrak{F}))^{-1} & , \text{ if } \xi = \vartheta \text{ or } \xi = (\eta_1,\eta_2,1) , \\ 0 & , \text{ if } \xi = (\eta_1,\eta_2,-1) . \end{cases}$$

Proof: Since e_j is Γ_{ξ}-invariant

$$g : z \longmapsto S_{\xi}^{(1)} e_j(M_{\xi}^{-1}(z))$$

is Γ_{ξ}'-invariant, i.e. λ_{ξ}-periodic in $u_{\xi}(z)$ by (3.8). It follows that the functions

$$g_n : z \longmapsto \frac{1}{\lambda_{\xi}} \int_0^{\lambda_{\xi}} e(-\frac{n}{\lambda_{\xi}}\varrho) g(\exp(\varrho X_{\xi})(z)) d\varrho$$

satisfy

$$g_n(\exp(\tau X_{\xi})(z)) = e(\frac{n}{\lambda_{\xi}}\tau) g_n(z) ,$$

where n is an integer and τ is real. Moreover by (2.9) and the G-invariance of Δ

$$\Delta g_n(z) + s_j(1-s_j) g_n(z) = 0$$

for z in $\mathfrak{H}$ and by (4.3)

$$T g_n(z) = \varepsilon g_n(z)$$

if $\xi = (\eta_1,\eta_2,\varepsilon)$. Therefore Lemma 3 and the remark following it yield numbers $\alpha_{j\xi}(n)$ and $\beta_{j\xi}(n)$ such that

$$g_n(z) = \begin{cases} \alpha_{j\xi}(0) y^{1/2} \log y + \beta_{j\xi}(0) V_{\xi}(z,1/2,0) & , \text{ if } \xi = \vartheta,\ s_j = 1/2 \text{ and } n = 0 , \\ \alpha_{j\xi}(n) U_{\xi}(z,s_j,\frac{n}{\lambda_{\xi}}) + \beta_{j\xi}(n) V_{\xi}(z,s_j,\frac{n}{\lambda_{\xi}}) & , \text{ otherwise}, \end{cases} \quad (5.5)$$

for all $j > 0$ and z in $\mathfrak{H}$. Since (2.9) and (4.9) hold for all z in $\mathfrak{H}$ both g_n and U_ξ are real-analytic on $\mathfrak{H}$. For $\xi = \zeta$ or η, on the other hand, V_ξ is not smooth on $\mathfrak{H}$ by (4.22), (4.25) and (4.27), whence $\beta_{j\xi}(n) = 0$ in those cases. To prove this also for $\xi = \vartheta$ we note that for sufficiently large X

$$\int_{\mathfrak{S}(X)} |g_n(z)|^2 d\omega(z) \leqslant \int_{\mathfrak{S}(X)} |g(z)|^2 d\omega(z) \leqslant \int_{\mathfrak{F}} |e_j(z)|^2 d\omega(z) = 1 , \quad (5.6)$$

where $\mathfrak{S}(X)$ consists of $z = x + iy$ with $0 < x < 1$ and $y > X$. In (5.6) we first used Bessel's inequality, then the fact that $\mathfrak{S}(X)$ is part of a fundamental domain for $M_\vartheta \Gamma M_\vartheta^{-1}$ if X is large enough (cf. Lemma 1.26 [22]) and finally the G-invariance of ω. Since by our choice $\mathrm{Re}\, s_j \geqslant 1/2$ it now follows from (2.3), (4.15), (4.16), (4.24), (4.25), (5.5) and (5.6) that $\beta_{j\vartheta}(n) = 0$ as well as that $\alpha_{j\vartheta}(0) = 0$ if $\mathrm{Re}\, s_j = 1/2$.

Since e_o is a constant function the corresponding g_n vanishes identically unless $n = 0$. Now $U_\zeta(z,1,0) = \gamma_\zeta(0,0) U_\zeta(z,0,0) = 2\pi$ by Lemma 4 and (4.7) while $U_\vartheta(z,1,0) = B(1/2,1/2) = \pi$ by (4.15). If $\xi = (\eta_1, \eta_2, 1)$ it follows from (4.1), (4.6), (4.7) and Cauchy's theorem that

$$U_\xi(z,1,0) = \int_{\mathcal{B}_\xi} (Y_\xi(\varrho,z) - Y_\xi(\varrho + i\pi, z))\, d\varrho = 2\pi i \, \mathrm{Res}_{\varrho = -\bar{w}_\xi} Y_\xi(\varrho,z) = 2\pi .$$

On the other hand, $S_\xi^{(1)} e_o(M_\xi^{-1}(z)) = 0$ for $\xi = (\eta_1, \eta_2, -1)$ by (4.1). Thus $\alpha_{o\xi}(0)$ has the desired form by (2.10) and (4.1).

Finally, Fourier series inversion yields

$$g(z) = \sum_{n \in Z} g_n(z) ,$$

where the sum converges absolutely and locally uniformly on $\mathfrak{H}$ since g is real-analytic. This proves the lemma.

Next we introduce exponential sums which will be related to the Fourier coefficients $\alpha_{j\xi}(n)$ in the main theorem. First we put

$$ {}_{\xi}\Gamma_{\chi} = \Gamma'_{\xi} \setminus M_{\xi} \Gamma M_{\chi}^{-1} \cap {}_{\xi}G_{\chi} / \Gamma'_{\chi} \tag{5.7} $$

$$ {}_{\xi}\gamma_{\chi} = \Gamma'_{\xi} \setminus M_{\xi} \Gamma M_{\chi}^{-1} \cap {}_{\xi}g_{\chi} / \Gamma'_{\chi} \tag{5.8} $$

and

$$ {}_{\xi}\sigma_{\chi} = \Gamma'_{\xi} \setminus M_{\xi} \Gamma M_{\chi}^{-1} \cap {}_{\xi}\mathfrak{s}_{\chi} / \Gamma'_{\chi} \tag{5.9} $$

where ${}_{\xi}G_{\chi}$, ${}_{\xi}g_{\chi}$ and ${}_{\xi}\mathfrak{s}_{\chi}$ are given by (3.9)-(3.12). In the notation of Lemma 1 we define generalized Kloosterman sums by

$$ {}_{\xi}^{\delta}S_{\chi}^{\delta'}(m,n,\nu) = \sum e\left(\frac{m}{\lambda_{\xi}}\, {}_{\xi}\Lambda_{\chi}^{\ell}(M) + \frac{n}{\lambda_{\chi}}\, {}_{\xi}\Lambda_{\chi}^{r}(M)\right) , \tag{5.10} $$

where the summation runs over a complete set of representatives M for ${}_{\xi}\Gamma_{\chi}$ such that ${}_{\xi}\nu_{\chi}(M) = \nu$, ${}_{\xi}\delta_{\chi}(M) = \delta$ and ${}_{\xi}\delta'_{\chi}(M) = \delta'$. Moreover we set

$$ {}_{\xi}s_{\chi}(m,n,\nu) = \sum e\left(\frac{m}{\lambda_{\xi}}\, {}_{\xi}\Lambda_{\chi}^{\ell}(M) + \frac{n}{\lambda_{\chi}}\, {}_{\xi}\Lambda_{\chi}^{r}(M)\right) \tag{5.11} $$

with summation over a complete set of representatives M for ${}_{\xi}\gamma_{\chi}$ such that

$$ M = \pm \exp({}_{\xi}\Lambda_{\chi}^{\ell}(M) X_{\xi}) \exp(\nu X_{\zeta}) \exp({}_{\xi}\Lambda_{\chi}^{r}(M) X_{\chi}) $$

and $|\nu - \pi| < \pi$. If the summation in (5.10) or (5.11) is over the empty set ${}_{\xi}^{\delta}S_{\chi}^{\delta'}(m,n,\nu)$ or ${}_{\xi}s_{\chi}(m,n,\nu)$ are understood to be zero.

Note that there are integers k, ℓ such that

$$ M = \pm \exp(k\lambda_{\xi} X_{\xi}) N \exp(\ell \lambda_{\chi} X_{\chi}) $$

if $\Gamma'_{\xi} M \Gamma'_{\chi} = \Gamma'_{\xi} N \Gamma'_{\chi}$. Thus the above sums are independent of the chosen representatives by Lemma 1. The following lemma ensures in particular that the sums in (5.10) and (5.11) are finite.

<u>Lemma 6</u>.

(i) For every $C > 0$ there are only finitely many double cosets $\Gamma'_{\xi} M \Gamma'_{\chi}$ in ${}_{\xi}\Gamma_{\chi}$ such that ${}_{\xi}\nu_{\chi}(M) \leq C$. In particular, these numbers

${}_\xi\nu_\chi(M)$ form a discrete subset of the positive real numbers.

(ii) There are only finitely many double cosets $\Gamma'_\xi M \Gamma'_\chi$ in ${}_\xi\gamma_\chi$. Moreover ${}_\xi\gamma_\chi$ can be non-empty only for $(\xi,\chi) = (\eta,\zeta')$, (ζ,η') or (η,η'). Then all representatives M of its double cosets satisfy ${}_\eta\nu_{\zeta'}(M) = \frac{\pi}{2}$, ${}_\zeta\nu_{\eta'}(M) = -\frac{\pi}{2}$ or $0 < |{}_\eta\nu_{\eta'}(M) - \pi| < \pi$ respectively.

(iii) In the notation of (2.4) and (2.6) ${}_\xi\sigma_\chi$ is non-empty if and only if $\xi \sim \chi$ or, in case $\xi = \eta$, $\eta^* \sim \chi$. More precisely the double cosets in ${}_\xi\sigma_\chi$ are

$$\begin{cases} \Gamma'_\xi \ , & \text{if } \xi \sim \chi \text{ and } \eta^* \not\sim \chi \text{ for } \xi = \eta, \\ \Gamma'_\xi M_\eta M_{\eta^*}^{-1} \ , & \text{if } \xi = \eta \not\sim \chi \text{ and } \eta^* \sim \chi \ , \\ \Gamma'_\xi \text{ and } \Gamma'_\xi M_\eta M_{\eta^*}^{-1} \ , & \text{if } \xi = \eta \sim \chi \text{ and } \eta^* \sim \chi \ , \\ \emptyset \ , & \text{otherwise,} \end{cases}$$

where $\xi \not\sim \chi$ means the negation of $\xi \sim \chi$.

<u>Proof</u>. First we show that the numbers ${}_\xi\nu_\chi(M)$ with $\Gamma'_\xi M \Gamma'_\chi$ in ${}_\xi\Gamma_\chi$ have a positive lower bound, say $\nu_o > 0$. This follows from Lemma 1(i) with $\nu_o = 1$ unless $\xi = \vartheta$ or $\chi = \vartheta'$. Since ${}_\xi\nu_\chi(M) = {}_\chi\nu_\xi(M^{-1})$ as we have seen in the proof of Lemma 1 it remains to consider the case $\xi = \vartheta$. Suppose, on the contrary, that there is a sequence $(M_n)_{n=1}^\infty$ with $\Gamma'_\vartheta M_n \Gamma'_\chi$ in ${}_\vartheta\Gamma_\chi$ such that ${}_\vartheta\nu_\chi(M_n) \to 0$ for $n \to \infty$. We may assume of course that $0 \leqslant {}_\vartheta\Lambda^r_\chi(M_n) \leqslant \lambda_\chi$ for all n. Therefore, if $\delta'_n = {}_\vartheta\delta'_\chi(M_n)$, all

$$R_\chi R_\vartheta^{1-\delta'_n} \exp({}_\vartheta\Lambda^r_\chi(M_n) X_\chi)(i) \ , \quad n = 1,2,\ldots \ ,$$

lie in a fixed compact set of $\mathfrak{H}$. Hence Lemma 1(i) implies that $\operatorname{Im} M_n(z) \to \infty$ for $n \to \infty$ since by assumption ${}_\vartheta\nu_\chi(M_n) \to 0$. This, however, means precisely that elements of Γ map $M_\chi^{-1}(i)$ into every horocyclic neighbourhood of the cusp ϑ . By Lemma 1.27[22] this cannot happen, whence $\nu_o > 0$.

By Lemma 1(i) the set of M in ${}_\xi G_\chi$ with $0 \leq {}_\xi\Lambda_\chi^\ell(M) \leq \lambda_\xi$, $0 \leq {}_\xi\Lambda_\chi^r(M) \leq \lambda_\chi$ and $\nu_o \leq {}_\xi\nu_\chi(M) \leq C$ is obviously compact for every $C > 0$. Since $M_\xi \Gamma M_\chi^{-1}$ is discrete and every double coset in ${}_\xi\Gamma_\chi$ has a representative M with $0 \leq {}_\xi\Lambda_\chi^\ell(M) \leq \lambda_\xi$ and $0 \leq {}_\xi\Lambda_\chi^r(M) \leq \lambda_\chi$ Lemma 6(i) now follows.

Similarly, Lemma 6(ii) is a consequence of (3.12), (3.16), Lemma 1 (ii) and Lemma 1(iii) since our tables in section 3 show that ${}_\xi\mathcal{S}_\chi = {}_\xi\mathcal{f}_\chi$ unless $(\xi,\chi) = (\eta,\zeta'), (\zeta,\eta')$ or (η,η') .

Finally, suppose that ${}_\xi\sigma_\chi$ is non-empty, i.e. there is an M in G such that $M(z)$ is in $\mathcal{f}_\xi$ for some z in $\mathcal{f}_\chi$ and $N = M_\xi^{-1} M M_\chi$ belongs to Γ . If $\xi = \zeta$ the first condition means that $M(i) = i$ and $\chi = \zeta'$, whence $N(\zeta') = M_\xi^{-1}(i) = \zeta$, i.e. $\zeta \sim \zeta'$. If $\xi = \vartheta$ the first condition means that either $M(\infty) = \infty$ and $\chi = \vartheta'$ or $M^{-1}(\infty) \in \{0,\infty\}$ and $\chi = (\eta_1', \eta_2', \varepsilon')$. Thus in the first case $N(\vartheta') = M_\xi^{-1}(\infty) = \vartheta$ i.e. $\vartheta \sim \vartheta'$ while in the second $N^{-1}(\vartheta) \in \{\eta_1', \eta_2'\}$ which means that ϑ is fixed by parabolic and hyperbolic elements in Γ . Hence the second case can never occur in view of Proposition 1.17 [22]. Eventually, if $\xi = \eta$ the first condition above implies that $\chi = \vartheta'$ or $\chi = \eta'$. For the same reason as before the case $\chi = \vartheta'$ is impossible now. Moreover it follows that either $M(0) = 0$ and $M(\infty) = \infty$, i.e. $\eta \sim \eta'$, or $M(0) = \infty$ and $M(\infty) = 0$, i.e. $\eta^* \sim \eta'$. For the fixpoint sets of two hyperbolic elements in Γ are identical or disjoint. Indeed, if N_1 , N_2 were two hyperbolic elements in Γ with just one common fixpoint then a contradiction to the discreteness of Γ would arise from the consideration of $N_1^k N_2 N_1^{-k}$, k in $\mathbb{Z}$.

On the other hand, if $\xi \sim \chi$ then $\Gamma'_\xi = \Gamma'_\chi$ is a double coset in ${}_\xi\sigma_\chi$ since $M_\xi \Gamma M_\chi^{-1} = M_\xi \Gamma M_\xi^{-1}$ by (5.4). If $\xi = \eta$ and $\eta^* \sim \chi$ then $\Gamma'_\xi M_\eta M_{\eta^*}^{-1} \Gamma'_\chi = \Gamma'_\xi M_\eta M_{\eta^*}^{-1}$ is another coset in ${}_\xi\sigma_\chi$ again by (5.4). These cosets exhaust ${}_\xi\sigma_\chi$ by what we noted about M in the preceding paragraph. Thus the lemma follows.

Remark 1. As an example let us consider the case $\Gamma = SL_2(\mathbb{Z})$ and $\xi = \chi = \infty$. Then $\lambda_\xi = \lambda_\chi = 1$ and in the notation of (2.1)

$$ {}_\xi\Lambda_\chi^\ell(M) = M(\infty) = \frac{a}{c} \quad , \quad {}_\xi\Lambda_\chi^r(M) = -M^{-1}(\infty) = \frac{d}{c} $$

$$ {}_\xi\nu_\chi(M) = |c| \quad , \quad {}_\xi\delta_\chi(M) = {}_\xi\delta'_\chi(M) = 0 $$

by section 3. The set of matrices with $c = \nu$, $0 \leq a < c$ and $0 \leq d < c$ form a complete set of representatives for ${}_\xi\Gamma_\chi$ with ${}_\xi\nu_\chi(M) = \nu$. For every double coset in ${}_\xi\Gamma_\chi$ has a representative M such that $c > 0$ and ${}_\xi\Lambda_\chi^\ell(M)$, ${}_\xi\Lambda_\chi^r(M)$ belong to the interval $[0,1)$. On the other hand, different matrices of our set obviously represent different double cosets. Thus

$$ {}_\infty^{\,o}S_\infty^{o}(m,n,\nu) = \sum_{\substack{0<a<c=\nu \\ ad\equiv 1 \,(\mathrm{mod}\ c)}} e\left(\frac{ma+nd}{c}\right) $$

since for $\Gamma = SL_2(\mathbb{Z})$ d is uniquely determined by a through the conditions $0 \leq d < c$ and $ad \equiv 1 \pmod c$. In other words, if $\Gamma = SL_2(\mathbb{Z})$ then ${}_\infty^{\,o}S_\infty^{o}(m,n,c)$ are the classical Kloosterman sums.

Remark 2. It is easy to deduce from Lemma 1 that ${}_\xi\gamma_\chi$ is non-empty if and only if there is a χ' with $\chi' \sim \chi$ such that χ' lies on $M_\xi^{-1}(\mathcal{F}_\xi)$, ξ lies on $M_{\chi'}^{-1}(\mathcal{F}_{\chi'})$ and $M_\xi^{-1}(\mathcal{F}_\xi)$ intersects $M_{\chi'}^{-1}(\mathcal{F}_{\chi'})$ for $(\xi,\chi) = (\eta,\xi')$, (ξ,η') and (η,η') respectively. This provides a geometric criterion for ${}_\xi\gamma_\chi$ being non-empty.

6. Computation of some integrals I

In this section we study the integral

$$J = S_{\chi}^{(1)} \int_{\mathfrak{p}_\chi} e(-\lambda'\varrho) V_{\xi}({}_{\xi}N_{\chi} \exp(\varrho X_{\chi})(z), s, \lambda)\, d\varrho \ , \tag{6.1}$$

with

$$ {}_{\xi}N_{\chi} = R_{\xi}^{-1} \begin{pmatrix} \nu & o \\ o & 1/\nu \end{pmatrix} R_{\chi} R_{\vartheta} \tag{6.2}$$

for $\nu \geqslant \nu_o > 0$. Later we shall choose ν_o as in the proof of Lemma 6 (i). Thus we assume from now on that $\nu_o > 1$ unless $\xi = \vartheta$ or $\chi = \vartheta'$. First we look for conditions under which J is defined by an absolutely convergent integral.

Let $\mathfrak{U}_{\chi}$ denote the set of z in $\mathfrak{H}$ such that

$$\begin{cases} v_{\chi}(z) > v_{\chi}(R_{\chi}^{-1}(i\nu_o^2)) \quad , \text{ if } \chi = \zeta' \text{ or } \vartheta' \ , \\ |\frac{\pi}{2} - v_{\chi}(z)| < \frac{\pi}{2} - v_{\chi}(R_{\chi}^{-1}(i\nu_o^2)) \ , \text{ if } \chi = \eta' \ . \end{cases} \tag{6.3}$$

Since $\mathfrak{U}_{\chi}$ is H_{χ}-invariant and V_{ξ} is real-analytic on $\mathfrak{H} - \mathfrak{F}_{\xi}$ the integrand of J is real-analytic if ${}_{\xi}N_{\chi}^{-1}(\mathfrak{F}_{\xi})$ is disjoint from $\mathfrak{U}_{\chi}$. By (3.5), (3.16) and (6.2) ${}_{\xi}N_{\chi}^{-1}(\mathfrak{F}_{\xi})$ consists of

$$\begin{cases} R_{\chi}^{-1}(\pm i\nu^2) \ , \text{ if } \xi = \zeta \ , \\ R_{\chi}^{-1}(\infty) \ , \text{ if } \xi = \vartheta \ , \\ \text{the points on the geodesic which ends in } R_{\chi}^{-1}(\pm\nu^2) \ , \text{ if } \xi = \eta \ . \end{cases}$$

Hence by (3.7) and elementary hyperbolic geometry ${}_{\xi}N_{\chi}^{-1}(\mathfrak{F}_{\xi})$ is indeed empty if $\nu \geqslant \nu_o$. It follows from (3.7), (3.16) and (4.20) that the integrand in (6.1) is

$$<< \begin{cases} 1 \ , \text{ if } \chi = \zeta' \\ v_{\xi}^{\sigma}(R_{\xi}^{-1}(\nu^2[z+\varrho])) << (1+\varrho^2)^{-\sigma} \ , \text{ if } \chi = \vartheta' \ , \\ v_{\xi}^{\sigma}\left(R_{\xi}^{-1}\left(\nu^2 \dfrac{ze^{\varrho}-1}{ze^{\varrho}+1}\right)\right) << e^{-\sigma|\varrho|} \ , \text{ if } \chi = \eta' \ , \end{cases}$$

uniformly for ϱ in $\mathbb{R}$ and locally uniformly for z in $\mathfrak{U}_{\chi}$. Thus the integral in (6.1) converges absolutely and locally uniformly for

z in $\mathfrak{U}_\chi$ and s in

$$\begin{cases} \mathbb{C} & , \text{ if } \chi = \zeta' , \\ \text{the half-plane } \sigma > 1/2 & , \text{ if } \chi = \vartheta' , \\ \text{the half-plane } \sigma > 0 & , \text{ if } \chi = \eta' . \end{cases} \tag{6.4}$$

Under these restrictions on z and s we shall compute J. Roughly speaking we want to proceed by inserting the integral representation of V_ξ and by interchanging the two integrals which so arise. In trying to do this we are confronted with two problems. First the path of integration in (4.8) for $V_\xi({}_\xi N_\chi \exp(\varrho X_\chi)(z),s,\lambda)$ depends on ϱ and secondly we have to justify the interchange. The first problem is overcome as follows.

Let $\mathfrak{D}'_\xi(z)$ denote a path in the τ-plane whose winding number with respect to the poles of $\tau \mapsto Y_\xi(\tau,z)$ is zero except for $\tau = -\bar{w}_\xi$ or $\tau = -w_\xi$ when this winding number is supposed to be 1 and -1 respectively. Here again w_ξ stands for $w_\xi(z)$. Since the poles of $Y_\xi(\tau,z)$ are simple there is a unique continuous determination of $\log Y_\xi(\tau,z)$ on $\mathfrak{D}'_\xi(z)$ such that

$$(1-e(s))i\, V_\xi(z,s,\lambda) = S_\xi^{(2)} \int_{\mathfrak{D}'_\xi(z)} e(-\lambda\tau)\, Y_\xi^{1-s}(\tau,z)\, d\tau , \tag{6.5}$$

where $Y_\xi^s(\tau,z)$ is given by $e(\frac{s}{2\pi i} \log Y_\xi(\tau,z))$. If, for instance, $\mathfrak{D}'_\xi(z)$ crosses $\mathfrak{D}_\xi(z)$ from the left to the right in just one point τ_o then $\log Y_\xi(\tau_o,z)$ is real for the branch chosen in (6.5).

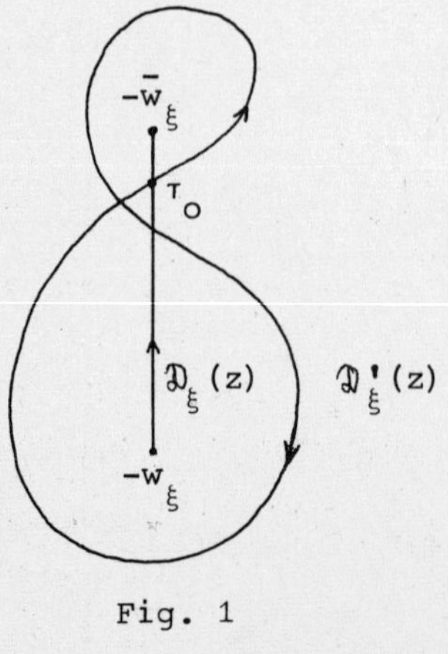

Fig. 1

Next we observe that by (3.2) and (6.2)

$$\varrho \longmapsto {}_{\xi}N_{\chi}\exp(\varrho X_{\chi})(z)$$

describe orthogonal trajectories of the geodesics

$$\begin{cases} \text{starting from } R_{\xi}^{-1}(\nu^2 \beta_{\chi}) \text{ , if } \chi = \zeta' \text{ or } \vartheta' \text{ ,} \\ \text{crossing the geodesic which ends in } R_{\xi}^{-1}(\pm\nu^2 \beta_{\chi}) \\ \text{orthogonally, if } \chi = \eta' \text{ ,} \end{cases}$$

where

$$\beta_{\chi} = \begin{cases} i \text{ , if } \chi = \zeta' \text{ ,} \\ \infty \text{ , if } \chi = \vartheta' \text{ ,} \\ 1 \text{ , if } \chi = \eta' \text{ .} \end{cases} \tag{6.6}$$

With this and (3.6) in mind we readily verify that for z and ν as assumed there always exists a path ${}_{\xi}\mathfrak{D}_{\chi}$ satisfying the defining properties of $\mathfrak{D}'_{\xi}({}_{\xi}N_{\chi}\exp(\varrho X_{\chi})(z))$ simultaneously for all ϱ in $\mathbb{R}$. More explicitly, for all real ϱ the winding number of ${}_{\xi}\mathfrak{D}_{\chi}$ with respect to

$$\tau \text{ or } \bar{\tau} \equiv -\bar{w}_{\xi}({}_{\xi}N_{\chi}\exp(\varrho X_{\chi})(z)) \pmod{\ell_{\xi}} \tag{6.7}$$

is zero except when τ or $\bar{\tau}$ are actually equal to the right hand side of (6.7). Then this winding number is 1 and -1 respectively. At least as far as the conditions on non-vanishing winding numbers are concerned the following pictures show paths qualifying for ${}_{\xi}\mathfrak{D}_{\chi}$:

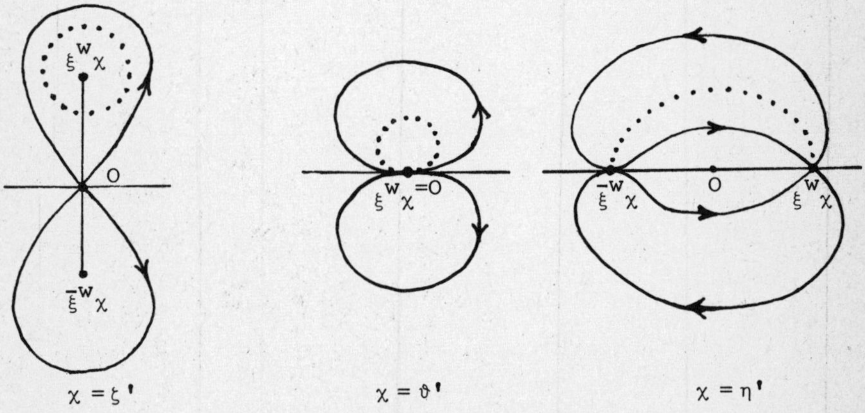

Fig. 2

In these pictures the dotted curves are parametrized by

$$\varrho \mapsto -\bar{w}_{\xi}({}_{\xi}N_{\chi}\exp(\varrho X_{\chi})(z))$$

and

$$ {}_{\xi}w_{\chi} = -\bar{w}_{\xi}(R_{\xi}^{-1}(\nu^2 \mathfrak{z}_{\chi})) \ . \tag{6.8}$$

From (4.2), (6.1) and (6.5) we obtain

$$(1-e(s))iJ = S_{\chi}^{(1)} \int_{\mathcal{P}_{\chi}} e(-\lambda'\varrho) \int_{{}_{\xi}\mathfrak{D}_{\chi}} e(-\lambda\tau) Y_{\xi}^{1-s}(\tau, {}_{\xi}N_{\chi}\exp(\varrho X_{\chi})(z))\, d\tau\, d\varrho \ , \tag{6.9}$$

where complex powers of Y_{ξ} are fixed as in (6.5). For we can omit $S_{\xi}^{(2)}$ on the right hand side of (6.9) since by our considerations following (6.3) ${}_{\eta}N_{\chi}\exp(\varrho X_{\chi})(z)$ has a positive real part for z in $\mathfrak{U}_{\chi}$. In (6.9) the contours no longer depend on the variables of integration and we now prepare grounds for the interchange of these two integrals.

By definition of Y_{ξ}, (4.4) and (6.2) we have

$$Y_{\xi}(\tau, {}_{\xi}N_{\chi}\exp(\varrho X_{\chi})(z)) = Y(M \exp(\varrho X_{\chi}), z) \ , \tag{6.10}$$

where

$$M = R_{\xi}\exp(\tau X_{\xi}) R_{\xi}^{-1} \begin{pmatrix} \nu & 0 \\ 0 & 1/\nu \end{pmatrix} R_{\chi} R_{\vartheta} \ . \tag{6.11}$$

We deduce from (3.4), (3.20) and (6.6) that the M above belongs to $\mathcal{S}_{\chi}$ precisely if

$$\exp(\tau X_{\xi}) R_{\xi}^{-1}(\pm \nu \mathfrak{z}_{\chi}) = R_{\xi}^{-1}(\infty) \ .$$

On applying w_{ξ} to both sides of this equation we deduce from (3.8), (3.16) and (6.8) that it is equivalent with

$$\tau \equiv \pm {}_{\xi}w_{\chi} \quad (\mathrm{mod}\, \ell_{\xi}) \ . \tag{6.12}$$

Therefore M decomposes according to Lemma 2 whenever the congruence (6.12) does not hold. On the other hand, if τ satisfies (6.12) and lies on ${}_{\xi}\mathfrak{D}_{\chi}$ then $\chi = \vartheta'$ or η' and $\tau = \pm {}_{\xi}w_{\chi}$ by the definition of these paths. Thus by Lemma 2 and (4.5) there is a non-zero complex number $\mu = \mu(M)$ such that

$$Y(M\exp(\varrho X_\chi),z) = \frac{1}{\mu^2(M)} Y_\chi(\varrho+\Lambda_\chi^r(M),z) \tag{6.13}$$

for all τ on ${}_\xi\mathfrak{D}_\chi$ except if $\tau = \pm\, {}_\xi w_\chi$.

Equation (6.13) will subsequently play an important rôle. First, however, we compute $\mu^2(M)$ and $\Lambda_\chi^r(M)$ explicitly in terms of τ, ν, ξ and χ. We deduce from (3.2) and (3.16) that

$$R_\xi \exp(\tau X_\xi) R_\xi^{-1} = \begin{pmatrix} * & * \\ c_\xi(\tau) & d_\xi(\tau) \end{pmatrix},$$

where

$$-c_\xi(\tau) = \begin{cases} \sin\frac{\tau}{2}, & \text{if } \xi = \zeta \\ \tau, & \text{if } \xi = \vartheta \\ \operatorname{sh}\frac{\tau}{2}, & \text{if } \xi = \eta \end{cases} \qquad d_\xi(\tau) = \begin{cases} \cos\frac{\tau}{2}, & \text{if } \xi = \zeta, \\ 1, & \text{if } \xi = \vartheta, \\ \operatorname{ch}\frac{\tau}{2}, & \text{if } \xi = \eta. \end{cases} \tag{6.14}$$

If we set ${}_\xi\Phi_\chi(\tau,\nu) = \Lambda_\chi^r(M)$ we therefore obtain from (3.21) and (6.11)

$${}_\xi\Phi_\chi(\tau,\nu) = -w_\chi\left(R_\vartheta R_\chi^{-1}\left(-\frac{d_\xi(\tau)}{c_\xi(\tau)}\right)\right). \tag{6.15}$$

Now

$$Y(M,R_\chi^{-1}(z)) = Y(R_\xi\exp(\tau X_\xi)R_\xi^{-1}, -\frac{\nu^2}{z})$$

by (3.16), (4.4) and (6.11) while

$$Y(M,R_\chi^{-1}(z)) = \frac{1}{\mu^2(M)} Y(R_\chi\exp({}_\xi\Phi_\chi(\tau,\nu)X_\chi)R_\chi^{-1},z)$$

by Lemma 2 and (4.4). On inserting the definition of Y on the right hand sides of these two equations we conclude that

$$\frac{y\nu^2/|z|^2}{(-c_\xi(\tau)\frac{\nu^2}{z}+d_\xi(\tau))(-c_\xi(\tau)\frac{\nu^2}{\bar z}+d_\xi(\tau))} =$$

$$= \frac{y}{\mu^2(M)(c_\chi({}_\xi\Phi_\chi(\tau,\nu))z+d_\chi({}_\xi\Phi_\chi(\tau,\nu)))(c_\chi({}_\xi\Phi_\chi(\tau,\nu))\bar z+d_\chi({}_\xi\Phi_\chi(\tau,\nu)))}$$

for all z in $\mathfrak{H}$. If we set ${}_\xi Q_\chi(\tau,\nu) = \mu^2(M)$ we thus obtain

$${}_\xi Q_\chi(\tau,\nu) = \left(\frac{d_\xi(\tau)}{\nu c_\chi({}_\xi\Phi_\chi(\tau,\nu))}\right)^2$$

by comparing coefficients on both sides of the preceding equation. This expression for ${}_{\xi}Q_{\chi}$ and (6.15) become more transparent when listed individually with respect to χ. By (3.6), (3.16) and (6.6) we have

$$
{}_{\xi}\Phi_{\chi}(\tau,\nu) = \begin{cases} -\dfrac{1}{\beta_{\chi}} \log\left(\dfrac{\nu^2 c_{\xi}(\tau) - \beta_{\chi} d_{\xi}(\tau)}{\nu^2 c_{\xi}(\tau) + \beta_{\chi} d_{\xi}(\tau)}\right) , & \text{if } \chi = \zeta' \text{ or } \eta' , \\ \dfrac{d_{\xi}(\tau)}{\nu^2 c_{\xi}(\tau)} , & \text{if } \chi = \vartheta' , \end{cases} \tag{6.16}
$$

whence by (6.14) and $2(\operatorname{sh}\frac{z}{2})^2 = \operatorname{ch} z - 1$

$$
{}_{\xi}Q_{\chi}(\tau,\nu) = \begin{cases} \left(\dfrac{\beta_{\chi} d_{\xi}(\tau)}{\nu \operatorname{sh}(\frac{1}{2}\beta_{\chi\,\xi}\Phi_{\chi}(\tau,\nu))}\right)^2 = (\nu c_{\xi}(\tau))^2 - \left(\dfrac{\beta_{\chi} d_{\xi}(\tau)}{\nu}\right)^2 , & \text{if } \chi = \zeta' \text{ or } \eta' , \\ \left(\dfrac{d_{\xi}(\tau)}{\nu\,{}_{\xi}\Phi_{\vartheta'}(\tau,\nu)}\right)^2 = (\nu c_{\xi}(\tau))^2 , & \text{if } \chi = \vartheta' . \end{cases} \tag{6.17}
$$

Now note that for ϱ in $\mathbb{R}$

$$
\tau \equiv \bar{w}_{\xi}({}_{\xi}N_{\chi} \exp(\varrho X_{\chi})(z)) \pmod{\ell_{\xi}} \tag{6.18}
$$

is equivalent with

$$
\exp(-\tau X_{\xi}) R_{\xi}^{-1}(\infty) = {}_{\xi}N_{\chi} \exp(\varrho X_{\chi})(\bar{z})
$$

by (3.6), (3.8) and (3.16). On applying ${}_{\xi}N_{\chi}^{-1}$ and w_{χ} to the preceding equation we infer from (3.6) and (3.8) that it is also equivalent with

$$
w_{\chi}({}_{\xi}N_{\chi}^{-1} \exp(-\tau X_{\xi})\; R_{\xi}^{-1}(\infty)) \equiv \varrho + w_{\chi}(\bar{z}) \pmod{\ell_{\chi}} .
$$

Thus we conclude from (6.2), (6.14) and (6.15) that (6.18) is equivalent with

$$
{}_{\xi}\Phi_{\chi}(\tau,\nu) \equiv -\bar{w}_{\chi}(z) - \varrho \pmod{\ell_{\chi}} \tag{6.19}
$$

for real ϱ. In view of (6.7) and (6.16) we can therefore choose ${}_{\xi}\mathfrak{D}_{\chi}$ in Fig. 2 in such a way that

$$
|\operatorname{Im}{}_{\xi}\Phi_{\chi}(\tau,\nu)| \leq \frac{1}{2} \operatorname{Im}(-\bar{w}_{\chi}(z)) \tag{6.20}
$$

and

$$\mathrm{Im}\, {}_{\xi}\Phi_{\chi}(\tau,\nu) \to 0 \;, \quad \tau \to 0 \;, \tag{6.21}$$

hold for τ on ${}_{\xi}\mathfrak{D}_{\chi}$ if $\chi = \zeta'$ or ϑ'. We also observe that ${}_{\xi}Q_{\chi}(\tau,\nu)$ and $Y_{\xi}(\varrho + {}_{\xi}\Phi_{\chi}(\tau,\nu), z)$ are regular and non-zero on ${}_{\xi}\mathfrak{D}_{\chi}$ except for $\tau = \pm\, {}_{\xi}w_{\chi}$ by (4.5), (6.6), (6.8), (6.16), (6.17) and our choice of ${}_{\xi}\mathfrak{D}_{\chi}$. Moreover those two expressions are positive for non-zero real τ if $\chi = \zeta'$ or ϑ'. If $\chi = \zeta'$ or ϑ' we thus obtain from (4.1), (6.9), (6.10) and (6.13)

$$(1-e(s))iJ = \int_{\mathcal{B}_{\chi}} e(-\lambda'\varrho) \int_{{}_{\xi}\mathfrak{D}_{\chi}} e(-\lambda\tau)\, {}_{\xi}Q_{\chi}^{s-1}(\tau,\nu) Y_{\chi}^{1-s}(\varrho + {}_{\xi}\Phi_{\chi}(\tau,\nu), z)\, d\tau d\varrho \;. \tag{6.22}$$

In (6.22) $\log {}_{\xi}Q_{\chi}(\tau,\nu)$ and $\log Y_{\chi}(\varrho + {}_{\xi}\Phi_{\chi}(\tau,\nu), z)$ are fixed continuously on ${}_{\xi}\mathfrak{D}_{\chi} - \{0\}$ such that their imaginary parts approach 0 as τ tends to 0 on the branch of ${}_{\xi}\mathfrak{D}_{\chi}$ in Fig. 2 which goes from the lower to the upper half-plane. For (6.10) is positive if τ is real whence its continuous logarithm used in (6.9) is real in $\tau = 0$ on the just mentioned branch of ${}_{\xi}\mathfrak{D}_{\chi}$.

If $\chi = \zeta'$ the contours in (6.22) are compact and the integrand is continuous on them. Thus interchange and substitution yield

$$(1-e(s))iJ = \int_{{}_{\xi}\mathfrak{D}_{\chi}} e(-\lambda\tau + \lambda'\, {}_{\xi}\Phi_{\chi}(\tau,\nu))\, {}_{\xi}Q_{\chi}^{s-1}(\tau,\nu) \int_{\mathcal{B}_{\chi} + {}_{\xi}\Phi_{\chi}(\tau,\nu)} e(-\lambda'\varrho) Y_{\chi}^{1-s}(\varrho, z)\, d\varrho d\tau$$

for $\chi = \zeta'$. Here the inner integral is a continuous function of τ on ${}_{\xi}\mathfrak{D}_{\chi}$ by (4.6) and (6.20). Moreover it is locally constant by Cauchy's theorem and it equals $U_{\chi}(z, 1-s, \lambda')$ for $\tau = 0$ by (4.7) and (6.21). Hence we have by Lemma 4

$$(1-e(s))iJ = \gamma_{\chi}(s,\lambda') U_{\chi}(z,s,\lambda') \int_{{}_{\xi}\mathfrak{D}_{\chi}} e(-\lambda\tau + \lambda'\, {}_{\xi}\Phi_{\chi}(\tau,\nu))\, {}_{\xi}Q_{\chi}^{s-1}(\tau,\nu)\, d\tau \tag{6.23}$$

for $\chi = \zeta'$, where complex powers are fixed as in (6.22).

Next we show that (6.23) also holds for $\chi = \vartheta'$ if $\lambda' \neq 0$ and $\sigma > 1/2$ although (6.22) is then no longer absolutely convergent. We deduce from (4.6), (6.14), (6.16), (6.17) and (6.20) that for $\chi = \vartheta'$

$$\int_{{}_\xi\mathfrak{D}_\chi} |{}_\xi Q_\chi(\tau,\nu)|^{\sigma-1} \int_{\mathcal{B}_\chi} |\frac{\partial}{\partial\varrho} Y_\chi^{1-s}(\varrho + {}_\xi\Phi_\chi(\tau,\nu), z)|\,|d\varrho|\,|d\tau| \ll$$

$$\ll \int_{{}_\xi\mathfrak{D}_\chi} |\tau|^{2(\sigma-1)} \int_{-\infty}^{\infty} \frac{d\varrho}{(1+|\varrho|)^{3-2\sigma}}\,|d\tau| \ll 1$$

locally uniformly for s in $\frac{1}{2} < \sigma < 1$ and z in $\mathfrak{U}_\chi$. By partial integration, interchange and substitution we thus obtain from (6.22)

$$(1-e(s))iJ = \int_{{}_\xi\mathfrak{D}_\chi} e(-\lambda\tau + \lambda'\,{}_\xi\Phi_\chi(\tau,\nu))\,{}_\xi Q_\chi^{s-1}(\tau,\nu) \int_{\mathcal{B}_\chi + {}_\xi\Phi_\chi(\tau,\nu)} e(-\lambda'\varrho)\,\frac{\partial}{\partial\varrho} Y_\chi^{1-s}(\varrho,z)\,\frac{d\varrho}{2\pi i\lambda'}\,d\tau$$

if $\chi = \vartheta'$, $\lambda' \neq 0$ and $\frac{1}{2} < \sigma < 1$. Now the inner integral equals

$$\int_{\mathcal{B}_\chi} e(-\lambda'\varrho)\,\frac{\partial}{\partial\varrho} Y_\chi^{1-s}(\varrho,z)\,\frac{d\varrho}{2\pi i\lambda'}$$

by an argument as in case of $\chi = \zeta'$. This latter integral is absolutely convergent for $\sigma < 1$. By undoing the partial integration and comparing with (4.7) we see that it equals $U_\chi(z,1-s,\lambda')$ in the half-plane $\sigma < 1/2$. Thus we conclude from Lemma 4 that (6.23) also holds for $\chi = \vartheta'$ if $\lambda' \neq 0$ and $1/2 < \sigma < 1$. Note that by (6.14), (6.17) and (6.21) the integral on the right of (6.23) is absolutely convergent for $\sigma > 1/2$ if $\chi = \vartheta'$.

Finally, if $\chi = \eta'$ we denote the two branches of ${}_\xi\mathfrak{D}_\chi$ starting from $\pm{}_\xi w_\chi$ and ending in $\mp{}_\xi w_\chi$ by ${}_\xi\mathfrak{D}_\chi^\pm$. Moreover we set

$$_\xi\Phi_\chi^\pm(\tau,\nu) = -\log\left(\pm\frac{\nu^2 c_\xi(\tau) - d_\xi(\tau)}{\nu^2 c_\xi(\tau) + d_\xi(\tau)}\right). \tag{6.24}$$

By (6.14) ${}_\xi\Phi_\chi^-(\tau,\nu)$ is real for $-{}_\xi w_\chi < \tau < {}_\xi w_\chi$ and ${}_\xi\Phi_\chi^+(\tau,\nu)$ is real if, modulo $\mathrm{Re}\,\ell_\xi$, τ does not lie in the interval $[-{}_\xi w_\chi, {}_\xi w_\chi]$. Our convention on logarithms yields

$$ {}_{\xi}\Phi_{\chi}^{+}(\tau,\nu) = {}_{\xi}\Phi_{\chi}^{-}(\tau,\nu) \pm i\pi $$

for $\pm \operatorname{Im}\tau > 0$, whence

$$ {}_{\xi}\Phi_{\chi}^{+}(\tau,\nu) \equiv -\bar{w}_{\chi}(z)-\varrho \pmod{\ell_{\chi}} \quad \text{and} \quad {}_{\xi}\Phi_{\chi}^{-}(\tau,\nu) \equiv -w_{\chi}(-\bar{z})-\varrho \pmod{\ell_{\chi}} $$

are equivalent conditions for ϱ in $\mathbb{R}$ and z in $\mathfrak{H}$. Hence by (6.7), (6.18) and (6.19) we may choose ${}_{\xi}\mathfrak{D}_{\chi}$ in Fig. 2 such that

$$ |\operatorname{Im} {}_{\xi}\Phi_{\chi}^{+}(\tau,\nu)| < \frac{1}{2} \operatorname{Im}(-\bar{w}_{\chi}(z)) \tag{6.25} $$

for τ on ${}_{\xi}\mathfrak{D}_{\chi}^{+}$,

$$ |\operatorname{Im}{}_{\xi}\Phi_{\chi}^{-}(\tau,\nu)| < \frac{1}{2} \operatorname{Im}(-\bar{w}_{\chi}(-\bar{z})) \tag{6.26} $$

for τ on ${}_{\xi}\mathfrak{D}_{\chi}^{-}$ and

$$ \operatorname{Im}{}_{\xi}\Phi_{\chi}^{\pm}(\tau,\nu) \to 0 \tag{6.27} $$

as τ tends to ${}_{\xi}^{-}w_{\chi}$ on ${}_{\xi}\mathfrak{D}_{\chi}^{\pm}$.

Now note that by (4.6), (6.14), (6.16) and (6.17)

$$ \pm{}_{\xi}Q_{\chi}(\tau,\nu) \quad \text{and} \quad \pm Y_{\chi}(\varrho+{}_{\xi}\Phi_{\chi}(\tau,\nu),z) $$

are positive if ϱ,τ and ${}_{\xi}\Phi_{\chi}^{\pm}(\tau,\nu)$ are real. Recall also that the logarithm of (6.10) used in (6.9) is real for $\tau = {}_{\xi}w_{\chi}$ on the branch of ${}_{\xi}\mathfrak{D}_{\chi}$ in Fig. 2 which leads from the lower to the upper half-plane. By (6.13) that logarithm therefore equals

$$ \log(\pm Y_{\chi}(\varrho+{}_{\xi}\Phi_{\chi}(\tau,\nu),z)) - \log(\pm{}_{\xi}Q_{\chi}(\tau,\nu)) $$

for τ on ${}_{\xi}\mathfrak{D}_{\chi}^{\pm}$, where these logarithms are fixed continuously on the two branches of ${}_{\xi}\mathfrak{D}_{\chi}^{\pm}$ such that

$$ \operatorname{Im}\log(\pm Y_{\chi}(\varrho+{}_{\xi}\Phi_{\chi}(\tau,\nu))) \to 0 $$

for $\tau \to {}_{\xi}w_{\chi}$ on ${}_{\xi}\mathfrak{D}^{\pm}_{\chi}$ and

$$\operatorname{Im}\log(\pm{}_{\xi}Q_{\chi}(\tau,\nu)) \to \begin{cases} 2\pi & , \text{ if } \tau \to {}_{\xi}w_{\chi} \text{ on } {}_{\xi}\mathfrak{D}^{\pm}_{\chi} \text{ in } \operatorname{Im}\tau > 0, \\ 0 & , \text{ if } \tau \to {}_{\xi}w_{\chi} \text{ on } {}_{\xi}\mathfrak{D}^{\pm}_{\chi} \text{ in } \operatorname{Im}\tau < 0. \end{cases} \tag{6.28}$$

With this determination of logarithms we now obtain from (6.9)

$$\begin{aligned}(1-e(s))iJ = S^{(1)}_{\chi}\int_{\mathcal{B}_{\chi}} e(-\lambda'\varrho)\{ &\int_{{}_{\xi}\mathfrak{D}^{+}_{\chi}} e(-\lambda\tau)\,{}_{\xi}Q^{s-1}_{\chi}(\tau,\nu)Y^{1-s}_{\chi}(\varrho+{}_{\xi}\Phi_{\chi}(\tau,\nu),z)\,d\tau \\ &+\int_{{}_{\xi}\mathfrak{D}^{-}_{\chi}} e(-\lambda\tau)(-{}_{\xi}Q_{\chi}(\tau,\nu))^{s-1}(-Y_{\chi}(\varrho+{}_{\xi}\Phi_{\chi}(\tau,\nu),z))^{1-s}d\tau\}d\varrho\end{aligned} \tag{6.29}$$

for $\chi = \eta'$. These integrals are absolutely convergent for $0<\sigma<1$ since by (4.6), (6.6), (6.8), (6.14), (6.17), (6.25) and (6.26)

$$\begin{aligned}\int_{{}_{\xi}\mathfrak{D}^{\pm}_{\chi}} |{}_{\xi}Q_{\chi}(\tau,\nu)|^{\sigma-1}\int_{\mathcal{B}_{\chi}} |Y_{\chi}(\varrho+{}_{\xi}\Phi_{\chi}(\tau,\nu),z)|^{1-\sigma}d\varrho\,|d\tau| &\ll \\ \ll \int_{{}_{\xi}\mathfrak{D}^{\pm}_{\chi}} |\tau^{2}-{}_{\xi}w^{2}_{\chi}|^{\sigma-1}\int_{\mathcal{B}_{\chi}} (\operatorname{ch}\varrho)^{\sigma-1}d\varrho\,|d\tau| &\ll 1\end{aligned}$$

if $0<\sigma<1$. In view of (4.6), (6.16) and (6.24) we have

$$-Y_{\chi}(\varrho+{}_{\xi}\Phi_{\chi}(\tau,\nu),z) = Y_{\chi}(\varrho+{}_{\xi}\Phi^{-}_{\chi}(\tau,\nu),-\bar{z})$$

for $\chi = \eta'$. Hence interchange and substitution transform (6.29) into

$$\begin{aligned}(1-e(s))iJ = &\int_{{}_{\xi}\mathfrak{D}^{+}_{\chi}} e(-\lambda\tau+\lambda'{}_{\xi}\Phi^{+}_{\chi}(\tau,\nu))\,{}_{\xi}Q^{s-1}_{\chi}(\tau,\nu)\,S^{(1)}_{\chi}\int_{\mathcal{B}_{\xi}+{}_{\xi}\Phi^{+}_{\chi}(\tau,\nu)} e(-\lambda'\varrho)Y^{1-s}_{\chi}(\varrho,z)\,d\varrho\,d\tau \\ &+\int_{{}_{\xi}\mathfrak{D}^{-}_{\chi}} e(-\lambda\tau+\lambda'{}_{\xi}\Phi^{-}_{\chi}(\tau,\nu))(-{}_{\xi}Q_{\chi}(\tau,\nu))^{s-1}S^{(1)}_{\chi}\int_{\mathcal{B}_{\chi}+{}_{\xi}\Phi^{-}_{\chi}(\tau,\nu)} e(-\lambda'\varrho)Y^{1-s}_{\chi}(\varrho,-\bar{z})\,d\varrho\,d\tau\,.\end{aligned}$$

By (6.25)-(6.27) the contours of the inner integrals above can once more be shifted to the real axis. Thus (4.1), (4.7) and Lemma 4 yield

$$(1-e(s))iJ = \gamma_\chi(s,\lambda')U_\chi(z,s,\lambda')\{\int_{{}_\xi\mathfrak{D}_\chi^+} e(-\lambda\tau+\lambda'\,{}_\xi\Phi_\chi^+(\tau,\nu))\,{}_\xi Q_\chi^{s-1}(\tau,\nu)\,d\tau$$

$$+\ \varepsilon' \int_{{}_\xi\mathfrak{D}_\chi^-} e(-\lambda\tau+\lambda'\,{}_\xi\Phi_\chi^-(\tau,\nu))\,(-{}_\xi Q_\chi(\tau,\nu))^{s-1}\,d\tau\} \tag{6.30}$$

for $\chi = (\eta_1', \eta_2', \varepsilon')$, where complex powers are fixed as in (6.29). The integrals on the right of (6.30) converge absolutely for $\sigma > 0$ by (6.14) and (6.17).

Summing up from (6.1), (6.23) and (6.30) we have proved

Lemma 7. If ${}_\xi N_\chi$ is given by (6.2) then

$$S_\chi^{(1)} \int_{\mathcal{b}_\chi} e(-\lambda'\varrho)V_\xi({}_\xi N_\chi \exp(\varrho X_\chi)(z),s,\lambda)\,d\varrho = {}_\xi J_\chi(\nu,s,\lambda,\lambda')U_\chi(z,s,\lambda') ,$$

where the integral on the left hand side converges absolutely and locally uniformly for z in $\mathfrak{U}_\chi$ given in (6.3) and s in the domains listed in (6.4). For s in these domains ${}_\xi J_\chi(\nu,s,\lambda,\lambda')$ has an absolutely convergent integral representation unless $\chi = \vartheta'$ and $\lambda' = 0$. More specifically, $(1-e(s))i\,{}_\xi J_\chi(\nu,s,\lambda,\lambda')U_\chi(z,s,\lambda')$ is equal to the right hand side of (6.23) if $\chi = \zeta'$ or $\chi = \vartheta'$ with $\lambda' \neq 0$, and to the right hand side of (6.30) if $\chi = \eta'$.

In the following lemma we take care of the still open case $\chi = \vartheta'$ and $\lambda' = 0$ by evaluating some ${}_\xi J_\chi$ explicitly. These ${}_\xi J_\chi$ will play a special rôle later on.

Lemma 8. If $\xi = \vartheta$, $\lambda = 0$ or $\chi = \vartheta'$, $\lambda' = 0$ we have

$${}_\xi J_\chi(\nu,s,\lambda,\lambda') = \beta_\xi B(\tfrac{1}{2},s)\,\nu^{-2s} ,$$

where β_ξ is given by (4.19) .

Proof: By (6.1) J depends continuously on λ' . Thus (4.7) Lemma 4 and (6.23) yield

$$(1-e(s))iJ = \pi^{1-2s} \frac{\Gamma(s)}{\Gamma(1-s)} U_\chi(z,s,o) \lim_{\lambda'\to o}\Bigl(|\lambda'|^{1-2s} \cdot$$

$$\int_{{}_\xi\mathcal{D}_\chi} e(-\lambda\tau+\lambda'\,{}_\xi\Phi_\chi(\tau,\nu))\,{}_\xi Q_\chi^{s-1}(\tau,\nu)\,d\tau\Bigr)$$

for $J = {}_\xi J_\chi(\nu,s,\lambda,o)$ if $\chi = \vartheta'$. It follows from (6.14), (6.16) and (6.17) that for $\chi = \vartheta'$

$$\lambda'\,{}_\xi\Phi_\chi(\beta_\xi\lambda'\tau,\nu) \to -\frac{1}{\nu^2\tau} \quad \text{and} \quad (\lambda')^{-2}\,{}_\xi Q_\chi(\beta_\xi\lambda'\tau,\nu) \to (\nu\tau)^2 ,$$

as $\lambda' \to 0$ locally uniformly in $\tau \neq 0$. Thus by Cauchy's theorem

$$ {}_\xi J_{\vartheta'}(\nu,s,\lambda,o) = \frac{\beta_\xi\pi^{1-2s}\Gamma(s)}{(1-e(s))i\Gamma(1-s)} \int_{{}_\vartheta\mathcal{D}_{\vartheta'}} e\Bigl(-\frac{1}{\nu^2\tau}\Bigr)(\nu\tau)^{2(s-1)}d\tau , \tag{6.31}$$

where complex powers of $(\nu\tau)^2 = {}_\vartheta Q_{\vartheta'}(\tau,\nu)$ are fixed as in (6.22). By contracting the upper half of ${}_\vartheta\mathcal{D}_{\vartheta'}$ to 0 and substituting $\varrho = -1/\nu^2\tau$ we conclude that the preceding integral equals

$$\nu^{-2s}\int_{\mathcal{C}_\vartheta+\varrho_o} e(\varrho)\varrho^{-2s}d\varrho ,$$

where $\operatorname{Im}\varrho_o < 0$ and $\log\varrho$ is fixed continuously on $\mathcal{C}_\vartheta+\varrho_o$ such that $-\pi < \operatorname{Im}\log\varrho < 0$. Since $\sigma > 1/2$ it is permissible to replace $\mathcal{C}_\vartheta+\varrho_o$ by a loop winding around the positive imaginary axis. On such a loop the integral converges locally uniformly for all s in $\mathbb{C}$. If $\sigma < 1/2$ we let the loop approach the imaginary axis and obtain

$$\int_{\mathcal{C}_\vartheta+\varrho_o} e(\varrho)\varrho^{-2s}d\varrho = i(e(-\tfrac{s}{2})-e(\tfrac{3s}{2}))\int_0^\infty e^{-2\pi\tau}\tau^{-2s}d\tau =$$

$$= i(e(-\tfrac{s}{2})-e(\tfrac{3s}{2}))\,\Gamma(1-2s)\,(2\pi)^{2s-1} .$$

Hence (6.31) and the functional equations for $\Gamma(s)$ yield

$${}_\xi J_{\vartheta'}(\nu,s,\lambda,o) = \frac{2^{2s-1}\Gamma(s)\Gamma(1-2s)}{\Gamma(1-s)\sin\pi s}\sin(2\pi s)\beta_\xi\nu^{-2s} = \beta_\xi B(1/2,s)\nu^{-2s}$$

as asserted.

For the second part we note that

$$ {}_{\vartheta}\Phi_{\chi}(\tau,\nu) = {}_{\vartheta}\Phi_{\chi}(\nu^2\tau,1) \quad \text{and} \quad {}_{\vartheta}Q_{\chi}(\tau,\nu) = \frac{1}{\nu^2}\, {}_{\vartheta}Q_{\chi}(\nu^2\tau,1) $$

by (6.14), (6.16) and (6.17), whence

$$ {}_{\vartheta}J_{\chi}(\nu,s,o,\lambda') = \nu^{-2s}\, {}_{\vartheta}J_{\chi}(1,s,o,\lambda') . $$

Thus it remains to show that ${}_{\vartheta}J_{\chi}(1,s,o,\lambda') = B(1/2,s)$.

If $\lambda' \neq 0$, as we may assume for $\chi = \vartheta'$,

$$ {}_{\vartheta}J_{\vartheta'}(1,s,o,\lambda') = \frac{\Gamma(s)\,(\pi\lambda')^{1-2s}}{(1-e(s))\, i\,\Gamma(1-s)} \int_{{}_{\vartheta}\mathfrak{D}_{\vartheta'}} e(-\lambda'\tau)\, \tau^{2(s-1)} d\tau $$

by Lemma 4, (6.14), (6.16), (6.17) and (6.23), where complex powers of τ are fixed as in (6.31). Therefore the first part of this proof gives

$$ {}_{\vartheta}J_{\vartheta'}(1,s,o,\lambda') = \frac{\Gamma(s)\,\pi^{1-2s}}{(1-e(s))\, i\,\Gamma(1-s)} \int_{\mathcal{C}_{\vartheta}+\varrho_0} e(\varrho)\, \varrho^{-2s} d\varrho = B(1/2,s) . $$

Again by Lemma 4, (6.14), (6.16), (6.17) and (6.23) we have

$$ {}_{\vartheta}J_{\vartheta'}(1,s,o,\lambda') = \frac{\gamma_{\xi'}(s,\lambda')}{(1-e(s))\, i} \int_{{}_{\vartheta}\mathfrak{D}_{\xi'}} \left(\frac{\tau+i}{\tau-i}\right)^{-2\pi\lambda'} (1+\tau^2)^{s-1} d\tau , $$

where $2\pi\lambda'$ is an integer. If $\sigma > 2\pi|\lambda'|$ we let ${}_{\vartheta}\mathfrak{D}_{\xi'}$ approach the line segment from $-i$ to i . By our fixation of logarithms the preceding integral then becomes

$$ i(1-e(s)) \int_{-1}^{1} \left(\frac{\varrho-1}{\varrho+1}\right)^{2\pi\lambda'} (1-\varrho^2)^{s-1} d\varrho = i(-1)^{2\pi\lambda'} (1-e(s))\, 2^{2s-1} B(s+2\pi\lambda', s-2\pi\lambda') . $$

Therefore Lemma 4 and the functional equations for $\Gamma(s)$ yield

$$ {}_{\vartheta}J_{\xi'}(1,s,o,\lambda') = \frac{\Gamma(s)\,\Gamma(1-s+2\pi|\lambda'|)\,\Gamma(s+2\pi\lambda')\,\Gamma(s-2\pi\lambda')}{\Gamma(1-s)\,\Gamma(s+2\pi|\lambda'|)\,\Gamma(2s)\, 2^{1-2s}} (-1)^{2\pi\lambda'} = B(1/2,s) . $$

Finally, if $\chi = (\eta_1', \eta_2', \varepsilon')$,

$$
{}_{\vartheta}J_{\chi}(1,s,o,\lambda') = \frac{\gamma_{\chi}(s,\lambda')}{(1-e(s))i}\left\{ \int\limits_{{}_{\xi}\mathfrak{D}_{\chi}^{+}} \left(\frac{\tau-1}{\tau+1}\right)^{2\pi i\lambda'} (\tau^2-1)\,d\tau + \right.
$$

$$
\left. + \varepsilon' \int\limits_{{}_{\xi}\mathfrak{D}_{\chi}^{-}} \left(\frac{1-\tau}{1+\tau}\right)^{2\pi i\lambda'} (1-\tau^2)^{s-1}\,d\tau \right\} \tag{6.32}
$$

by (6.14), (6.16), (6.17), (6.24) and (6.30). If $0<\sigma<1/2$ we replace ${}_{\vartheta}\mathfrak{D}_{\chi}^{+}$ in (6.32) by a path of the following shape

Fig. 3

If this path approaches $(-\infty,-1)\cup(1,\infty)$ from above and below the first integral in (6.32) equals

$$
(1+e(s))\int\limits_{1}^{\infty}\left\{\left(\frac{\tau-1}{\tau+1}\right)^{2\pi i\lambda'} + \left(\frac{\tau-1}{\tau+1}\right)^{2\pi i\lambda'}\right\}(\tau^2-1)^{s-1}\,d\tau =
$$

$$
= (1+e(s))\,2^{2s-1}\{B(s+2\pi i\lambda',1-2s) + B(s-2\pi i\lambda',1-2s)\}
$$

by (6.27) and (6.28). Similarly the second integral in (6.32) can be written as

$$
(1+e(s))\int\limits_{-1}^{1}\left(\frac{1-\tau}{1+\tau}\right)^{2\pi i\lambda'}(1-\tau^2)^{s-1}\,d\tau = (1+e(s))\,2^{2s-1}B(s+2\pi i\lambda',s-2\pi i\lambda')
$$

if we shrink ${}_{\vartheta}\mathfrak{D}_{\chi}^{-}$ to the interval $(-1,1)$. Hence Lemma 4 and the functional equations for $\Gamma(s)$ imply that

$$
{}_{\vartheta}J_{\chi}(1,s,o,\lambda') = \gamma_{\chi}(s,\lambda')\,2^{2s-1}\operatorname{ctg}\pi s\; B(s+2\pi i\lambda',s-2\pi i\lambda')\left\{\frac{\cos(2\pi^2 i\lambda')}{\cos\pi s} + \varepsilon'\right\}
$$

$$
= \frac{\Gamma^2(s)\,2^{2s-2}(\cos(2\pi^2 i\lambda')+\varepsilon'\cos\pi s)}{\Gamma(2s)\sin\pi\left(\frac{1+\varepsilon'+2s}{4}+i\pi\lambda'\right)\sin\pi\left(\frac{1+\varepsilon'+2s}{4}-i\pi\lambda'\right)} = B(1/2,s)
$$

for $\chi = (\eta_1', \eta_2', \varepsilon')$. This completes the proof of Lemma 8.

Lemma 9. As a function of the second variable ${}_{\xi}J_{\chi}(\nu,s,\lambda,\lambda')$ is analytic in $\sigma > 0$ and satisfies

$$ {}_{\xi}J_{\chi}(\nu,s,\lambda,\lambda') = \beta_{\xi}B(1/2,s)\,\nu^{-2s} + O\left(\nu^{-2\sigma}\left[\frac{1}{\nu^{2}} + \frac{|s|}{|s|+\nu^{4}}\right]\right) $$

uniformly for $\nu \geqslant \nu_o$ and s in $\sigma_o \leqslant \sigma \leqslant 2$, where $\nu_o > 0$ and $\sigma_o > 0$.

Proof: If $\chi = \zeta'$ or η' the assertion on the analyticity follows from Lemma 7. For such a χ and given $\nu > 0$ we define a positive number $\nu' = {}_{\xi}\nu'_{\chi}$ by

$$ \nu^{2}\,\frac{c_{\xi}(\nu'\mathfrak{z}_{\chi})}{d_{\xi}(\nu'\mathfrak{z}_{\chi})} = -\,\mathfrak{z}_{\chi}\,. $$

Then we successively verify with the help of Taylor expansions, (6.6), (6.14), (6.16) and (6.17) that

$$ \nu' = \frac{\beta_{\xi}}{\nu^{2}}(1+O(\nu^{-4}))\,,\quad d_{\xi}(\varrho\nu') = 1+O(\nu^{-4})\,,\quad \left(\frac{c_{\xi}}{d_{\xi}}\right)^{(1)}(\mp\nu'\mathfrak{z}_{\chi}) = -\frac{1}{\beta_{\xi}} + O(\nu^{-4})\,, $$

$$ \nu^{2}\,\frac{c_{\xi}(\varrho\nu')}{d_{\xi}(\varrho\nu')} \pm \mathfrak{z}_{\chi} = \nu^{2}\nu'(\varrho\mp\mathfrak{z}_{\chi})\left\{\left(\frac{c_{\xi}}{d_{\xi}}\right)^{(1)}(\mp\mathfrak{z}_{\chi}\nu') + O(\nu^{-4})\right\} = $$

$$ -(\varrho\mp\mathfrak{z}_{\chi})(1+O(\nu^{-4}))\,. $$

$$ {}_{\xi}\Phi_{\chi}(\varrho\nu',\nu) = {}_{\vartheta}\Phi_{\chi}(\varrho,1) + O(\nu^{-4}) \quad\text{and}\quad {}_{\xi}Q_{\chi}(\varrho\nu',\nu) = \frac{1}{\nu^{2}}\,{}_{\vartheta}Q_{\chi}(\varrho,1)(1+O(\nu^{-4})) $$

as $\nu \to \infty$ uniformly for $|\varrho| << 1$. Thus the integral on the right hand side of (6.23) equals

$$ \nu'\int_{{}_{\vartheta}\mathfrak{D}_{\chi}} e(\lambda'\,{}_{\vartheta}\Phi_{\chi}(\varrho,1))(\nu^{-2}\,{}_{\vartheta}Q_{\chi}(\varrho,1))^{s-1}d\varrho + \nu'\int_{{}_{\xi}\mathfrak{D}_{\chi}} e(\lambda'\,{}_{\vartheta}\Phi_{\chi}(\varrho,1))\,{}_{\xi}Q_{\chi}^{s-1}(\varrho\nu',\nu) $$

$$ \{1+O(\nu^{-2}) - (1+O(\nu^{-4}))^{s-1}\}d\varrho $$

for $\chi = \zeta'$. Since by our choice $0 \leqslant \mathrm{Im}\log {}_{\xi}Q_{\chi}(\tau,\nu) \leqslant 2\pi$ on ${}_{\xi}\mathfrak{D}_{\chi}$ it therefore follows from Lemma 4, 7 and 8 that

$$ {}_{\xi}J_{\zeta'}(\nu,s,\lambda,\lambda') = \beta_{\xi}\nu^{-2s}\,{}_{\vartheta}J_{\zeta'}(1,s,o,\lambda') + O(\nu^{-2\sigma}|\gamma_{\zeta'}(s,\lambda')|[\nu^{-2}+\frac{|s|}{|s|+\nu^4}]) $$

$$ = \beta_{\xi}B(1/2,s)\,\nu^{-2s} + O\left(\nu^{-2\sigma}\left[\nu^{-2}+\frac{|s|}{|s|+\nu^4}\right]\right) $$

uniformly for ν and s in the desired ranges. The same reasoning implies the Lemma for $\chi = \eta'$ if we use (6.30) instead of (6.23).

Finally, by Lemma 8 it remains to consider $\chi = \vartheta'$ and $\lambda' \neq 0$. Lemma 7 shows that ${}_{\xi}J_{\vartheta'}(\nu,s,\lambda,\lambda')$ is analytic in $\sigma > 1/2$. We extend this range by partial integration. Since

$$ \frac{\partial}{\partial\tau}\, e(\lambda'\,{}_{\xi}\Phi_{\vartheta'}(\tau,\nu)) = \frac{2\pi i\lambda' e(\lambda'\,{}_{\xi}\Phi_{\vartheta'}(\tau,\nu))}{\beta_{\xi}\;{}_{\xi}Q_{\vartheta'}(\tau,\nu)} \tag{6.33} $$

and

$$ \frac{\partial}{\partial\tau}\,{}_{\xi}Q^{s}_{\vartheta'}(\tau,\nu) = -2s\,\frac{d_{\xi}(\tau)}{\beta_{\xi}c_{\xi}(\tau)}\;{}_{\xi}Q^{s}_{\vartheta'}(\tau,\nu) \tag{6.34} $$

by (6.14), (6.16) and (6.17) the integral on the right hand side of (6.23) equals

$$ \int_{{}_{\xi}\mathfrak{D}_{\chi}} e(-\lambda\tau+\lambda'\,{}_{\xi}\Phi_{\chi}(\tau,\nu))\,{}_{\xi}Q^{s}_{\chi}(\tau,\nu)\left[\frac{\beta_{\xi}\lambda}{\lambda'} + \frac{sd_{\xi}(\tau)}{\pi i\lambda' c_{\xi}(\tau)}\right] d\tau $$

for $\chi = \vartheta'$. This integral converges absolutely and locally uniformly for s in $\sigma > 0$ whence ${}_{\xi}J_{\vartheta'}(\nu,s,\lambda,\lambda')$ is analytic in $\sigma > 0$. Similarly as above we obtain with $\nu' = \beta_{\xi}/\nu^2$

$$ {}_{\xi}\Phi_{\vartheta'}(\varrho\nu',\nu) = {}_{\vartheta}\Phi_{\vartheta'}(\varrho,1) + O(\varrho/\nu^4) $$

and

$$ {}_{\xi}Q_{\vartheta'}(\varrho\nu',\nu) = \nu^{-2}\,{}_{\vartheta}Q_{\vartheta'}(\varrho,1)(1+O(\varrho^2/\nu^4)) $$

as $\nu \to \infty$ uniformly for $0 < |\varrho| \ll 1$. Therefore the above integral is

$$ -\frac{\beta_{\xi}s}{\pi i\lambda'} \int_{{}_{\vartheta}\mathfrak{D}_{\chi}} e(\lambda'\,{}_{\vartheta}\Phi_{\chi}(\varrho,1))(\nu^{-2}\,{}_{\vartheta}Q_{\chi}(\varrho,1))^{s}\,\frac{d\varrho}{\varrho} + $$

$$ \nu' \int_{{}_{\xi}\mathfrak{D}_{\chi}} e(\lambda'\,{}_{\vartheta}\Phi_{\chi}(\varrho,1))\,{}_{\xi}Q^{s}_{\chi}(\varrho\nu',\nu)\{O(1) - \frac{s\nu^2}{\varrho}[1+O(\varrho/\nu^2)-(1+O(\varrho^2/\nu^4))^{s}]\}d\varrho . $$

We undo the partial integration in the first of these integrals and note for the second that again $0 \leqslant \mathrm{Im}\log {}_{\xi}Q_{\chi}(\tau,\nu) \leqslant 2\pi$ on ${}_{\xi}\mathfrak{D}_{\chi}$. Thus Lemma 4, 7 and 8 yield

$$
{}_{\xi}J_{\vartheta'}(\nu,s,\lambda,\lambda') = \beta_{\xi}\;{}_{\vartheta}J_{\vartheta'}(\nu,s,o,\lambda') + O\Big(\nu^{-2\sigma-2}|s|^{2\sigma-1} + \big(\tfrac{|s|}{\nu}\big)^{2\sigma}\int_{{}_{\xi}\mathfrak{D}_{\vartheta'}} |\varrho|^{2\sigma-1} \tag{6.35}
$$

$$
|1+O(\varrho/\nu^{2}) - (1+O(\varrho^{2}/\nu^{4}))^{s}|\,|d\varrho| = \beta_{\xi}B(1/2,s)\,\nu^{-2s} + O\left(\big(\tfrac{|s|}{\nu}\big)^{2\sigma}\left[\nu^{-2} + \frac{|s|}{|s|+\nu^{4}}\left(\frac{\nu^{4}}{|s|+\nu^{4}}\right)^{\sigma}\right]\right)
$$

uniformly for $\nu \geqslant \nu_{o}$ and s in $0 \leqslant \sigma \leqslant 2$. By a different partial integration the integral on the right hand side of (6.23) also equals

$$
\frac{1}{1-s}\int_{{}_{\xi}\mathfrak{D}_{\chi}} e(-\lambda\tau+\lambda'\,{}_{\xi}\Phi_{\chi}(\tau,\nu))\,{}_{\xi}Q_{\chi}^{s-1}(\tau,\nu)\,\frac{c_{\xi}(\tau)}{d_{\xi}(\tau)}\left[\pi i\lambda\beta_{\xi} - \frac{i\pi\lambda'}{{}_{\xi}Q_{\chi}(\tau,\nu)} + \frac{1}{2c_{\xi}(\tau)d_{\xi}(\tau)}\right]d\tau
$$

for $\chi = \vartheta'$, where this integral converges absolutely for $\sigma > 1$. We approximate it by

$$
\frac{\beta_{\xi}}{1-s}\int_{{}_{\vartheta}\mathfrak{D}_{\chi}} e(\lambda'\,{}_{\vartheta}\Phi_{\chi}(\tau,\nu))\,{}_{\vartheta}Q_{\chi}^{s-1}(\tau,\nu)\left[\frac{i\pi\lambda'\tau}{{}_{\vartheta}Q_{\chi}(\tau,\nu)} + \frac{1}{2}\right]d\tau
$$

and thus obtain similarly as in (6.35)

$$
{}_{\xi}J_{\vartheta'}(\nu,s,\lambda,\lambda') = \beta_{\xi}\,{}_{\vartheta}J_{\vartheta'}(\nu,s,o,\lambda') + O\Big(\nu^{-2\sigma-2}|s|^{2\sigma-2} + \big(\tfrac{|s|}{\nu}\big)^{2\sigma}|s|^{-2}\int_{{}_{\xi}\mathfrak{D}_{\vartheta'}}|\varrho|^{2\sigma-3}
$$

$$
|1+O(\varrho/\nu^{2}) - (1+O(\varrho^{2}/\nu^{4}))^{s}|\,|d\varrho| = \beta_{\xi}B(1/2,s)\,\nu^{-2s} + O\left(\big(\tfrac{|s|}{\nu}\big)^{2\sigma}|s|^{-2}\left[\nu^{-2} + \frac{|s|}{|s|+\nu^{4}}\left(\frac{\nu^{4}}{\nu^{4}+|s|}\right)^{\sigma-1}\right]\right)
$$

uniformly for $\nu \geqslant \nu_{o}$ and s in $1 \leqslant \sigma \leqslant 2$. Hence (6.35) and the Phragmén-Lindelöf principle yield

$$ {}_{\xi}J_{\vartheta'}(\nu,s,\lambda,\lambda') = \beta_{\xi}B(1/2,s)\nu^{-2s} + O\left(\nu^{-2\sigma}\left[\nu^{-2}+\frac{|s|}{|s|+\nu^{4}}\right]\right) $$

uniformly for $\nu \geqslant \nu_o$ and s in $0 \leqslant \sigma \leqslant 1$. To complete the proof of Lemma 9 it remains to extend the range for s to $1 \leqslant \sigma \leqslant 2$ in the preceding formula. This is readily achieved by repeating the second type of partial integration, i.e. by integrating ${}_{\xi}Q_{\chi}^{s-1}(\tau,\nu)$.

<u>Remark</u>: Subsequently ε and ε' will always denote the third component of ξ and χ respectively if they have one. Nevertheless we shall write down sums of the form

$$ \sum_{\delta,\delta'=o}^{1} \varepsilon^{\delta}\varepsilon'^{\delta'}\,{}_{\xi}A_{\chi}(\delta,\delta') $$

for every pair (ξ,χ). In such a case a closer look will always reveal that ${}_{\xi}A_{\chi}(1,\delta') = 0$ unless $\xi = \eta$ and ${}_{\xi}A_{\chi}(\delta,1) = 0$ unless $\chi = \eta'$. Therefore the above sum is meaningful for every pair (ξ,χ) if ε^{o} and $(\varepsilon')^{o}$ are understood to be 1 whereever they occur.

<u>Lemma 10</u>. Let

$$ {}_{\xi}I_{\chi}(\nu,s,\lambda,\lambda') = \gamma_{\chi}(s,\lambda') \sum_{\delta,\delta'=o}^{1} \varepsilon^{\delta}\varepsilon'^{\delta'} \int_{{}_{\xi}\mathcal{C}_{\chi}^{\delta\delta'}} e(-\lambda\tau+\lambda'\mathrm{Re}\,{}_{\xi}\Phi_{\chi}(\tau+\delta\pi i,\nu)) \, |{}_{\xi}Q_{\chi}(\tau+\delta\pi i,\nu)|^{s-1}d\tau , $$

where ${}_{\xi}\mathcal{C}_{\chi}^{\delta\delta'}$ denotes that part of $\mathcal{C}_{\xi}$ on which $(-1)^{\delta+\delta'}\,{}_{\xi}Q_{\chi}(\tau+\delta\pi i,\nu)$ is positive. Then the above integrals converge locally uniformly for s in $0<\sigma<1$ unless $\xi = \vartheta$, $\lambda = 0$ or $\chi = \vartheta'$, $\lambda' = 0$. The convergence is absolute for s in the strip

$$ \begin{cases} 0<\sigma<1 , & \text{unless } \xi = \vartheta \text{ or } \chi = \vartheta' , \\ 0<\sigma<1/2 , & \text{if } \xi = \vartheta \text{ but not } \chi = \vartheta' \\ \frac{1}{2}<\sigma<1 , & \text{if } \chi = \vartheta' \text{ but not } \xi = \vartheta . \end{cases} $$

Moreover the following functional equations hold

$$ {}_{\xi}J_{\chi}(\nu,s,\lambda,\lambda') = \gamma_{\xi}(s,\lambda)\gamma_{\chi}(s,\lambda')\,{}_{\xi}J_{\chi}(\nu,1-s,\lambda,\lambda') + \operatorname{ctg}\pi s\,{}_{\xi}I_{\chi}(\nu,s,\lambda,\lambda') $$

and

$$ {}_{\xi}I_{\chi}(\nu,s,\lambda,\lambda') = \gamma_{\xi}(s,\lambda)\gamma_{\chi}(s,\lambda')\,{}_{\xi}I_{\chi}(\nu,1-s,\lambda,\lambda') , $$

unless $\xi = \vartheta,\ \lambda = 0$ or $\chi = \vartheta'$, $\lambda' = 0$.

<u>Proof</u>: To begin with we suppose that $\lambda\lambda' \neq 0$. As for (4.20) we deduce from (4.6) and (4.8) that

$$ \frac{\partial}{\partial\varrho} V_{\xi}({}_{\xi}N_{\vartheta'}\exp(\varrho X_{\vartheta'})(z),s,\lambda) \ll |\varrho|^{-2\sigma-1} , \quad |\varrho|\to\infty . $$

Therefore Lemma 4 and 7 yield

$$ \begin{aligned} &{}_{\xi}J_{\chi}(\nu,s,\lambda,\lambda')\, U_{\chi}(z,s,\lambda') - \gamma_{\xi}(s,\lambda)\,{}_{\xi}J_{\chi}(\nu,1-s,\lambda,\lambda')U_{\chi}(z,1-s,\lambda') \\ &= \operatorname{ctg}\pi s\ S_{\chi}^{(1)} \int_{\mathcal{B}_{\chi}} \frac{e(-\lambda'\varrho)}{2\pi i\lambda'} \frac{\partial}{\partial\varrho} U_{\xi}({}_{\xi}N_{\chi}\exp(\varrho X_{\chi})(z),1-s,\lambda)\,d\varrho , \end{aligned} \tag{6.36} $$

where the integral on the right converges absolutely in $0<\sigma<1$. Now let w denote a C^{∞}-function of compact support on $\mathbb{R}$ such that $w(\tau) = 1$ for $|\tau| \leqslant 10$. Since by (6.14) and (6.17)

$$ {}^{\delta}_{\xi}\mathcal{B}^{0}_{\chi} + {}^{\delta}_{\xi}\mathcal{B}^{1}_{\chi} = \begin{cases} \mathcal{B}_{\xi} , & \text{if } \delta = 0 \text{ or if } \delta = 1 \text{ and } \xi = \eta, \\ \emptyset , & \text{if } \delta = 1 \text{ and } \xi = \zeta \text{ or } \vartheta , \end{cases} $$

up to sets of measure 0 it follows from (4.1) and (4.7) that

$$ \begin{aligned} U_{\xi}(z,s,\lambda) = \sum_{\delta,\delta'=0}^{1} \varepsilon^{\delta} \int_{{}^{\delta}_{\xi}\mathcal{B}^{\delta'}_{\chi}} e(-\lambda\tau)\Big[w(\tau)\,((-1)^{\delta}Y_{\xi}(\tau+\delta\pi i,z))^{s} + \\ + \frac{1}{2\pi i\lambda}\frac{\partial}{\partial\tau}\Big\{(1-w(\tau))\,((-1)^{\delta}Y_{\xi}(\tau+\delta\pi i,z))^{s}\Big\}\Big]\,d\tau , \end{aligned} \tag{6.37} $$

where the integrals converge absolutely in $\sigma > 0$. For we note that

$$ TY_{\eta}(\tau,z) = -Y_{\eta}(\tau+\pi i,z) $$

by (4.6) . This last equation, (6.10) and (6.13-(6.17) also show that

$$(-1)^{\delta} Y_{\xi}(\tau+\delta\pi i, {}_{\xi}N_{\chi}\exp(\varrho X_{\chi})(z)) = \frac{Y_{\chi}(\varrho + {}_{\xi}\Phi_{\chi}(\tau+\delta\pi i,\nu), z)}{(-1)^{\delta}\,{}_{\xi}Q_{\chi}(\tau+\delta\pi i,\nu)} =$$

$$= \frac{T^{\delta'} Y_{\chi}(\varrho + \mathrm{Re}\,{}_{\xi}\Phi_{\chi}(\tau+\delta\pi i,\nu), z)}{(-1)^{\delta+\delta'}\,{}_{\xi}Q_{\chi}(\tau+\delta\pi i,\nu)}$$

if ϱ is on $\mathscr{C}_{\chi}$ and τ on ${}_{\xi}^{\delta}\mathscr{C}_{\chi}^{\delta'}$, since $\mathrm{Im}\,{}_{\xi}\Phi_{\chi}(\tau+\delta\pi i,\nu) \equiv \delta'\pi \pmod{2\pi}$ for τ on ${}_{\xi}^{\delta}\mathscr{C}_{\chi}^{\delta'}$. Thus an interchange of integrations and a simple substitution imply that the right hand side of (6.36) equals

$$\mathrm{ctg}\pi s \sum_{\delta,\delta'=0}^{1} \varepsilon^{\delta}\varepsilon'^{\delta'} \int_{{}_{\xi}^{\delta}\mathscr{C}_{\chi}^{\delta'}} e(-\lambda\tau)\,[\mathscr{W}(\tau)\,e(\lambda'\mathrm{Re}\,{}_{\xi}\Phi_{\chi}(\tau+\delta\pi i,\nu))\,|{}_{\xi}Q_{\chi}(\tau+\delta\pi i,\nu)|^{s-1} +$$

$$\frac{1}{2\pi i\lambda}\frac{\partial}{\partial\tau}\{(1-\mathscr{W}(\tau))\,e(\lambda'\mathrm{Re}\,{}_{\xi}\Phi_{\chi}(\tau+\delta\pi i,\nu))\,|{}_{\xi}Q_{\chi}(\tau+\delta\pi i,\nu)|^{s-1}\}]\,d\tau$$

$$\cdot\gamma_{\chi}(s,\lambda')\,U_{\chi}(z,s,\lambda') \ . \tag{6.38}$$

For by (4.3), (4.7), (4.13) and Lemma 4

$$S_{\chi}^{(1)} T^{\delta'} \int_{\mathscr{C}_{\chi}} \frac{e(-\lambda'\varrho)}{2\pi i\lambda'} \frac{\partial}{\partial\varrho} Y_{\chi}^{1-s}(\varrho,z)\,d\varrho = T^{\delta'} S_{\chi}^{(1)} \int_{\mathscr{C}_{\chi}} e(-\lambda'\varrho)\,Y_{\chi}^{1-s}(\varrho,z)\,d\varrho =$$

$$= T^{\delta'} U_{\chi}(z,1-s,\lambda') = \varepsilon'^{\delta'}\gamma_{\chi}(s,\lambda')\,U_{\chi}(z,s,\lambda')$$

if $\sigma < 1/2$.

To justify the foregoing interchange we note that the first integral in the preceding chain of equations converges absolutely in $\sigma < 1$ by (4.6). Then we readily verify with (4.19), (6.6) and (6.14) - (6.17) that

$$\frac{\partial}{\partial\tau}\,{}_{\xi}\Phi_{\chi}(\tau,\nu) = \frac{\beta_{\chi}}{\beta_{\xi}\,{}_{\xi}Q_{\chi}(\tau,\nu)} \quad\text{and}\quad \frac{\partial}{\partial\tau}\,{}_{\xi}Q_{\chi}(\tau,\nu) = \frac{2}{\beta_{\xi}}(-\nu^{2}+(\nu\mathfrak{z}_{\xi}\mathfrak{z}_{\chi})^{-2})\,c_{\xi}(\tau)\,d_{\xi}(\tau) \ . \tag{6.39}$$

Therefore the integrals in (6.38) converge absolutely for s in $\frac{1}{2} < \sigma < 1$ and we conclude that in this strip the interchange was permissible.

Since for $0 < \sigma < 1$ the curly brackets in (6.38) tend to 0 as $|\tau| \to \infty$ we can get rid of the derivative there on integrating by parts. It follows that (6.38) is equal to

$$\operatorname{ctg}\pi s \, {}_{\xi}I_{\chi}(\nu,s,\lambda,\lambda')U_{\chi}(z,s,\lambda')$$

if $\frac{1}{2} < \sigma < 1$. Thus we obtain the functional equation connecting ${}_{\xi}I_{\chi}$ and ${}_{\xi}J_{\chi}$ from Lemma 4 and (3.36). Moreover a partial integration of the first summand under the integral sign in (6.38) together with (6.39) yield

$$\begin{aligned}{}_{\xi}I_{\chi}(\nu,s,\lambda,\lambda') = \gamma_{\chi}(s,\lambda') \sum_{\delta,\delta'=0}^{1} \varepsilon^{\delta}\varepsilon'^{\delta'} \int_{\delta_{\xi}\mathcal{B}_{\chi}^{\delta'}} \Big[& e(\lambda' \operatorname{Re} {}_{\xi}\Phi_{\chi}(\tau+\delta\pi i,\nu)) \\ & \frac{(-1)^{\delta+\delta'+1}\beta_{\xi}}{2\pi i\lambda'\beta_{\chi}} \frac{\partial}{\partial\tau}\left\{ \mathcal{W}(\tau)e(-\lambda\tau)\,|{}_{\xi}Q_{\chi}(\tau+\delta\pi i,\nu)|^{s}\right\} + \\ & \frac{e(-\lambda\tau)}{2\pi i\lambda} \frac{\partial}{\partial\tau}\left\{(1-\mathcal{W}(\tau))e(\lambda'\operatorname{Re}{}_{\xi}\Phi_{\chi}(\tau+\delta\pi i,\nu))\,|{}_{\xi}Q_{\chi}(\tau+\delta\pi i,\nu)|^{s-1}\right\}\Big] d\tau \,,\end{aligned} \tag{6.40}$$

where now the integrals on the right converge absolutely in $0 < \sigma < 1$. We also observe that for $0 < \sigma < 1$ the first curly brackets tend to 0 as τ tends to a root of ${}_{\xi}Q_{\chi}(\tau+\delta\pi i,\nu)$. Hence we easily obtain all assertions on the defining integral representation of ${}_{\xi}I_{\chi}$ by undoing the partial integrations in (6.40).

Looking back we note that the partial integrations in (6.36) and (6.37) are required to ensure the absolute convergence of the double integral leading to (6.38) only if $\chi = \vartheta'$ and $\xi = \vartheta$ respectively. However, if these partial integrations are not introduced no restrictions on λ and λ' arise. Thus, if $\lambda\lambda' = 0$, we may assume that either $\xi = \vartheta$, $\chi = \zeta'$ or η' with $\lambda \neq 0$, $\lambda' = 0$ or $\chi = \vartheta'$, $\xi = \zeta$ or η with $\lambda' \neq 0$, $\lambda = 0$. In the first of these two cases the left hand side of (6.36) still equals (6.38) which then converges absolutely for s in $0 < \sigma < 1$. In the second case we do not use partial integration on the right of (6.37). Nevertheless the so arising double integral for the left hand side of (6.36) still converges absolutely in $\frac{1}{2} < \sigma < 1$. Moreover, by undoing the partial integration in the second summand under the integral sign in (6.40) we obtain an integral representation of ${}_{\xi}I_{\chi}$ converging absolutely for s in $0 < \sigma < 1$ if $\xi = \zeta$ or η. These facts again imply the functional equation between ${}_{\xi}I_{\chi}$ and ${}_{\xi}J_{\chi}$ as well as the asserted properties of convergence in the two cases mentioned above.

Finally, by Lemma 4

$$\gamma_{\xi}(s,\lambda)\,\gamma_{\xi}(1-s,\lambda) = 1$$

unless $\xi = \vartheta$, $\lambda = 0$. Thus a double application of the functional equation already obtained yields

$$\begin{aligned} \operatorname{ctg}\pi s\,{}_{\xi}I_{\chi}(\nu,s,\lambda,\lambda') &= \gamma_{\xi}(s,\lambda)\,\gamma_{\chi}(s,\lambda')\{\gamma_{\xi}(1-s,\lambda)\,\gamma_{\chi}(1-s,\lambda')\,{}_{\xi}J_{\chi}(\nu,s,\lambda,\lambda') - \\ &- {}_{\xi}J_{\chi}(\nu,1-s,\lambda,\lambda')\} = \gamma_{\xi}(s,\lambda)\,\gamma_{\chi}(s,\lambda')\operatorname{ctg}\pi s\,{}_{\xi}I_{\chi}(\nu,1-s,\lambda,\lambda') \end{aligned}$$

unless $\xi = \vartheta$, $\lambda = 0$ or $\chi = \vartheta'$, $\lambda' = 0$. This completes the proof of the lemma.

Lemma 10 is connected with the decomposition of ${}_{\xi}G_{\chi}$. In the same way the next lemma is related to the decomposition of ${}_{\xi}g_{\chi}$. For its formulation we introduce

$$
{}_\xi\varphi_\chi(\tau,\nu) = \begin{cases} -\frac{1}{i}\log\left(\frac{d_\xi(\tau)+ic_\xi(\tau)}{d_\xi(\tau)-ic_\xi(\tau)}\, e^{-i(\nu+r_\xi)}\right), & \text{if } \chi=\xi', \\ -\log\left(\frac{d_\xi(\tau)\cos\frac{\nu+r_\xi}{2}+c_\xi(\tau)\sin\frac{\nu+r_\xi}{2}}{d_\xi(\tau)\sin\frac{\nu+r_\xi}{2}-c_\xi(\tau)\cos\frac{\nu+r_\xi}{2}}\right), & \text{if } \chi=\eta', \end{cases} \tag{6.41}
$$

and

$$
{}_\xi q_\chi(\tau,\nu) = \begin{cases} c_\xi^2(\tau)+d_\xi^2(\tau), & \text{if } \chi=\xi', \\ 2\left(d_\xi(\tau)\cos\frac{\nu+r_\xi}{2}+c_\xi(\tau)\sin\frac{\nu+r_\xi}{2}\right)\left(d_\xi(\tau)\sin\frac{\nu+r_\xi}{2}-\right. \\ \qquad \left. -\,c_\xi(\tau)\cos\frac{\nu+r_\xi}{2}\right), & \text{if } \chi=\eta', \end{cases} \tag{6.42}
$$

where r_ξ is as in (3.16).

<u>Lemma 11</u>. If $(\xi,\chi) = (\eta,\xi')$, (ξ,η') or (η,η') and $0<\nu<2\pi$ with $\nu \neq \pi$ let

$$
{}_\xi i_\chi(\nu,s,\lambda,\lambda') = \gamma_\chi(s,\lambda') \sum_{\delta,\delta'=0}^{1} \varepsilon^\delta \varepsilon'^{\delta'} \int_{{}_\xi^\delta\mathcal{B}_\chi^{\delta'}} e(-\lambda\tau+\lambda'\operatorname{Re}{}_\xi\varphi_\chi(\tau+\delta\pi i,\nu))
$$
$$
|{}_\xi q_\chi(\tau+\delta\pi i,\nu)|^{s-1} d\tau ,
$$

where ${}_\xi^\delta\mathcal{B}_\chi^{\delta'}$ denotes that part of $\mathcal{B}_\xi$ on which $(-1)^{\delta+\delta'}{}_\xi q_\chi(\tau+\delta\pi i,\nu)$ is positive. Then

$$
S_\chi^{(1)} \int_{\mathcal{B}_\chi} e(-\lambda'\varrho)\, U_\xi(\exp(\nu X_\zeta)\exp(\varrho X_\chi)(z),1-s,\lambda)\, d\varrho = {}_\xi i_\chi(\nu,s,\lambda,\lambda')\, U_\chi(z,s,\lambda') .
$$

All integrals above converge absolutely and locally uniformly for s in $0<\sigma<1$. Moreover ${}_\xi i_\chi$ is meromorphic in s and satisfies

$$
{}_\xi i_\chi(\nu,s,\lambda,\lambda') = \gamma_\xi(s,\lambda)\,\gamma_\chi(s,\lambda')\,{}_\xi i_\chi(\nu,1-s,\lambda,\lambda') .
$$

<u>Proof</u>: Let (ξ,χ) denote one of the three pairs the lemma deals with and $M = R_\xi \exp(\tau X_\xi)\exp(\nu X_\zeta)$. Lemma 2 and (4.5) show that

$$
Y(M,z) = \frac{1}{\mu^2}\, Y_\chi(\Lambda_\chi^r(M),z)
$$

unless M is in $\mathcal{T}_\chi$. Here μ^2 and $\Lambda_\chi^r(M)$ can explicitly be computed in the following way. First (3.6), (3.16), (3.21), (6.14) and (6.41) lead to

$$\Lambda^r_\chi(M) = -w_\chi(M^{-1}(\infty)) = -w_\chi(\exp(-[\nu+r_\xi]X_\zeta)R_\xi\exp(-\tau X_\xi)R_\xi^{-1}(\infty))$$

$$= -w_\chi\left(\frac{d_\xi(\tau)\cos\frac{\nu+r_\xi}{2}+c_\xi(\tau)\sin\frac{\nu+r_\xi}{2}}{d_\xi(\tau)\sin\frac{\nu+r_\xi}{2}-c_\xi(\tau)\cos\frac{\nu+r_\xi}{2}}\right) = {}_\xi\varphi_\chi(\tau,\nu) \quad .$$

Secondly by (3.16) and (4.4)

$$Y(M,R_\chi^{-1}(z)) = Y(R_\xi\exp(\tau X_\xi)R_\xi^{-1},\exp([\nu+r_\xi-r_\chi]X_\zeta)(z))$$

$$= \frac{1}{\mu^2}\, Y(R_\chi\exp(\Lambda^r_\chi(M)X_\chi)R_\chi^{-1},z) \quad ,$$

whence the comparison of coefficients as after (6.15) yields

$$\mu^2 = \frac{\left(c_\xi(\tau)\cos\frac{\nu+r_\xi-r_\chi}{2} - d_\xi(\tau)\sin\frac{\nu+r_\xi-r_\chi}{2}\right)^2}{c_\chi^2({}_\xi\varphi_\chi(\tau,\nu))} \quad .$$

Thus it follows from (3.16), (6.14), (6.41), (6.42) and $2\,\mathrm{sh}^2 z = \mathrm{ch} z - 1$ that for $\chi = \xi'$

$$\mu^2 = \frac{-4\left[c_\xi(\tau)\cos\frac{\nu+r_\xi}{2} - d_\xi(\tau)\sin\frac{\nu+r_\xi}{2}\right]^2 {}_\xi q_\chi(\tau,\nu)}{\left[(d_\xi(\tau)+ic_\xi(\tau))e^{-i(\nu+r_\xi)/2} - (d_\xi(\tau)-ic_\xi(\tau))e^{i(\nu+r_\xi)/2}\right]^2} = {}_\xi q_\chi(\tau,\nu)$$

and for $\chi = \eta'$

$$\mu^2 = \frac{2\left[c_\xi(\tau)\cos\frac{\nu+r_\xi-r_\chi}{2} - d_\xi(\tau)\sin\frac{\nu+r_\xi-r_\chi}{2}\right]^2 {}_\xi q_\chi(\tau,\nu)}{\left[c_\xi(\tau)\left(\sin\frac{\nu+r_\xi}{2}+\cos\frac{\nu+r_\xi}{2}\right) - d_\xi(\tau)\left(\sin\frac{\nu+r_\xi}{2}-\cos\frac{\nu+r_\xi}{2}\right)\right]^2} = {}_\xi q_\chi(\tau,\nu).$$

Furthermore (3.4) and (3.20) show that M belongs to $\mathcal{S}_\chi$ precisely if

$$\begin{cases} \exp([\nu+r_\xi]X_\zeta)(\pm i) = \pm i = R_\xi \exp(-\tau X_\xi) R_\xi^{-1}(\infty) = -\dfrac{d_\xi(\tau)}{c_\xi(\tau)} \;, \text{ if } \chi = \zeta' \;, \\ \exp(-[\nu+r_\xi]X_\zeta)\left(-\dfrac{d_\xi(\tau)}{c_\xi(\tau)}\right) = 0 \text{ or } \infty \;, \text{ if } \chi = \eta', \end{cases}$$

which always turns out to be equivalent with the vanishing of ${}_\xi q_\chi(\tau,\nu)$ by (6.42).

Since up to sets of measure zero

$${}^\delta_\xi\mathcal{B}^0_\chi + {}^\delta_\xi\mathcal{B}^1_\chi = \begin{cases} \mathcal{b}_\xi \;, \text{ if } \delta = 0 \text{ or if } \delta = 1 \text{ and } \xi = \eta \;, \\ \emptyset \;, \text{ if } \delta = 1 \text{ and } \xi = \zeta \;, \end{cases}$$

we conclude from the above as in the proof of Lemma 10 that

$$U_\xi(z,s,\lambda) = \sum_{\delta,\delta'=0}^{1} \varepsilon^\delta \int_{{}^\delta_\xi\mathcal{B}^{\delta'}_\chi} e(-\lambda\tau)((-1)^\delta Y_\xi(\tau+\delta\pi i,z))^s d\tau \;, \; \sigma > 0 \;,$$

and that

$$(-1)^\delta Y_\xi(\tau+\delta\pi i, \exp(\nu X_\zeta)(z)) = (-1)^\delta Y(R_\xi \exp([\tau+\delta\pi i]X_\xi)\exp(\nu X_\zeta), z)$$

$$= \frac{Y_\chi({}_\xi\varphi_\chi(\tau+\delta\pi i,\nu),z)}{(-1)^\delta\, {}_\xi q_\chi(\tau+\delta\pi i,\nu)} = \frac{T^{\delta'} Y_\chi(\mathrm{Re}\,{}_\xi\varphi_\chi(\tau+\delta\pi i,\nu),z)}{(-1)^{\delta+\delta'}\, {}_\xi q_\chi(\tau+\delta\pi i,\nu)}$$

if τ is on ${}^\delta_\xi\mathcal{B}^{\delta'}_\chi$. For again ${}^\delta_\xi\mathcal{B}^1_\chi = \emptyset$ unless $\chi = \eta'$. Moreover we infer from (4.6), (6.14) and (6.42) that

$$\int_{\mathcal{b}_\chi} \int_{{}^\delta_\xi\mathcal{B}^{\delta'}_\chi} |Y_\xi(\tau+\delta\pi i, \exp(\nu X_\zeta)\exp(\varrho X_\chi)(z))|^{1-\sigma} d\tau d\varrho \ll$$

$$\ll \int_{\mathcal{b}_\chi} Y_\chi^{1-\sigma}(\varrho,z)\,d\varrho \int_{\mathcal{b}_\xi} |{}_\xi q_\chi(\tau+\delta\pi i,\nu)|^{\sigma-1} d\tau < \infty$$

if $0 < \sigma < 1$ and $0 < |\nu-\pi| < \pi$. Thus interchange, substitution, (4.3), (4.13) and Lemma 4 yield

$$S_\chi^{(1)} \int_{\mathcal{C}_\chi} e(-\lambda'\varrho) U_\xi(\exp(\nu X_\xi)\exp(\varrho X_\chi)(z), 1-s, \lambda)\, d\varrho =$$

$$= \sum_{\delta,\delta'=0}^{1} \varepsilon^\delta \int_{{}_\delta{}_\xi\mathcal{C}_\chi^{\delta'}} e(-\lambda\tau + \lambda' \mathrm{Re}\, {}_\xi\varphi_\chi(\tau+\delta\pi i, \nu)) \qquad (6.43)$$

$$|{}_\xi q_\chi(\tau+\delta\pi i, \nu)|^{s-1} d\tau T^{\delta'} U_\chi(z, 1-s, \lambda') = {}_\xi i_\chi(\nu, s, \lambda, \lambda') U_\chi(z, s, \lambda') .$$

By (6.14) and (6.42) the integral representation of ${}_\xi i_\chi$ converges absolutely and locally uniformly if $0 < \sigma < 1$ and $0 < |\nu - \pi| < \pi$. In (6.43) we now insert the functional equations for U_ξ and U_χ given in Lemma 4. As a result we obtain the desired functional equation for ${}_\xi i_\chi$.

By (6.14) and (6.42) the integration over a relatively compact neighbourhood of $\{\tau \mid {}_\xi q_\chi(\tau+\delta\pi i, \nu) = 0\}$ contributes to the integral representation of ${}_\xi i_\chi$ something analytic in $\sigma > 0$ while the contribution from the complement of that neighbourhood is analytic in $\sigma < 1$. Since the first mentioned contribution to ${}_\xi i_\chi$ extends meromorphically into $\sigma \leq 0$ by a standard use of power series expansions, the lemma now follows.

<u>Remark</u>: The functions ${}_\xi J_\chi$, ${}_\xi I_\chi$ and ${}_\xi i_\chi$ have a group representational significance. They allow us to describe explicitly how the eigenfunctions U_ξ and V_ξ transform under the regular representation of G on $\mathfrak{H}$. For example we obtain from Lemma 1 and 7

$$V_\xi(M(z), s, \lambda) = e(\lambda\, {}_\xi\Lambda_\chi^{\ell}(M)) \int_{\mathcal{C}_\chi} e(\lambda'\, {}_\xi\Lambda_\chi^{r}(M))\, {}_\xi J_\chi({}_\xi\nu_\chi(M), s, \lambda, \lambda') U_\chi(z, s, \lambda')\, d\lambda'$$

for all z in $\mathfrak{H}$ if M belongs to ${}_\xi G_\chi$ and $\xi = \vartheta$, $\chi = \vartheta'$.

7. Poincaré series and their Fourier series expansions

We how introduce Poincaré series including the Eisenstein series of section 2 and the resolvent kernel of the Laplacian as special cases. They are all Γ-invariant eigenfunctions of Δ and have a Fourier series expansion with respect to any χ. Unlike the situation we met in Lemma 5 these expansions usually converge only on a part of $\mathfrak{H}$ and V_χ may also occur in it. The main object of this section is to express the resulting Fourier coefficients explicitly by series involving the Kloosteman sums of section 5.

If A, C are subgroups and B is a right A- or left C-invariant subset of G then as usual

$$\sum_{M \in A\backslash B} \qquad \sum_{M \in B/C} \qquad \text{or} \qquad \sum_{M \in A\backslash B/C}$$

indicate summation over a complete set of representatives M for $A\backslash B$, B/C or $A\backslash B/C$ respectively. For every ξ and every integer m we define a Poincaré series by

$$P_\xi(z,s,m) = \sum_{M \in \Gamma_\xi \backslash \Gamma} V_\xi(z_{M_\xi M}, s, \frac{m}{\lambda_\xi}) \tag{7.1}$$

with V_ξ given by (4.8) and M_ξ, λ_ξ as in section 5. By (4.12) and the definition of λ_ξ the summands in (7.1) are independent of the chosen representatives since m is an integer. We also introduce

$$P_\xi^\psi(z,s,m) = \sum_{M \in \Gamma_\xi \backslash \Gamma} V_\xi(z_{M_\xi M}, s, \frac{m}{\lambda_\xi}) \psi(v_\xi(z_{M_\xi M})) , \tag{7.2}$$

where $v_\xi(z)$ is given by (3.7) and ψ is a non-negative C^∞-function on $\mathbb{R}$ such that

$$\psi(\varrho) = \begin{cases} \psi_o(\varrho) , & \text{if } \xi = \zeta \text{ or } \vartheta , \\ \psi_o(\varrho) + \psi_o(\pi - \varrho) , & \text{if } \xi = \eta , \end{cases} \quad \text{and} \quad \psi_o(\varrho) = \begin{cases} 1 , & \text{if } \varrho \leqslant 1 , \\ 0 , & \text{if } \varrho \geqslant 3/2 . \end{cases} \tag{7.3}$$

First we show

<u>Lemma 12</u>. The series (7.2) converge absolutely and locally uniformly for z in $\mathfrak{H}$ and s in the half-plane $\sigma > 1$. Moreover $P_\xi^\psi(z,s,m)$ is bounded on $\mathfrak{H}$ if $\sigma > 1$.

<u>Proof</u>: By (3.6) we have

$$\left|\frac{dz}{dw_\zeta}\right| = \frac{2e^{-v_\zeta}}{\left|1-e^{iw_\zeta}\right|^2} , \quad \left|\frac{dz}{dw_\eta}\right| = e^{u_\eta}$$

and

$$y = \frac{1-e^{-2v_\zeta}}{\left|1-e^{iw_\zeta}\right|^2} = e^{u_\eta}\sin v_\eta .$$

Thus we obtain from (2.3)

$$d\omega(z) = \left|\frac{dz}{dw_\xi}\right| \frac{du_\xi dv_\xi}{y^2} = \frac{du_\xi dv_\xi}{\omega_\xi^2(v_\xi)} , \tag{7.4}$$

where $du_\xi dv_\xi$ stands for integration with respect to the Lebesgue measure in the (u_ξ, v_ξ)-plane and

$$\omega_\xi(v) = \begin{cases} \mathrm{sh}v , & \text{if } \xi = \zeta , \\ v , & \text{if } \xi = \vartheta , \\ \sin v , & \text{if } \xi = \eta . \end{cases} \tag{7.5}$$

Let $\mathfrak{S}_\xi$ consist of $u+iv$ with

$$0 < u < \lambda_\xi \quad \text{and} \quad \begin{cases} 0 < v < \infty , & \text{if } \xi = \zeta \text{ or } \vartheta , \\ 0 < v < \pi , & \text{if } \xi = \eta . \end{cases} \tag{7.6}$$

Without further saying we often identify $\mathfrak{S}_\xi$ with the set of points z in $\mathfrak{H}$ for which $w_\xi(z)$ lies in $\mathfrak{S}_\xi$. Then we deduce from (4.20), (7.3) and the G-invariance of ω that

$$\int_{\mathfrak{F}} \sum_{M\in\Gamma_\xi\backslash\Gamma} v_\xi(z_{M_\xi M},\sigma,0)\,\psi_0(v_\xi(z_{M_\xi M}))\,d\omega(z) \ll \sum_{M\in\Gamma_\xi'\backslash M_\xi \Gamma M_\xi^{-1}} \tag{7.7}$$

$$\int_{M_\xi(\mathfrak{F})} v_\xi^\sigma(z_M)\,\psi_0(v_\xi(z_M))\,d\omega(z) = \int_{\mathfrak{S}_\xi} v^\sigma \psi_0(v)\,\frac{du\,dv}{\omega_\xi^2(v)} ,$$

since $M_\xi(\mathfrak{F})$ and $\mathfrak{F}_\xi$ are fundamental domains for $M_\xi \Gamma M_\xi^{-1}$ and Γ'_ξ respectively. By (7.3), (7.5) and (7.6) these integrals converge for $\sigma > 1$.

If $d(z,z_o)$ denotes the hyperbolic distance from z to z_o then by elementary hyperbolic geometry $d(z,z_o) \geqslant d(z',z_o)$, where z' is determined by the equations $u_\xi(z') = u_\xi(z_o)$ and $v_\xi(z') = v_\xi(z)$. Since in the various coordinate systems the hyperbolic metric is given by

$$ds^2 = \frac{dx^2+dy^2}{y^2} = \left|\frac{dz}{dw_\xi}\right|^2 \frac{du_\xi^2+dv_\xi^2}{y^2} = \frac{du_\xi^2+dv_\xi^2}{\omega_\xi^2(v_\xi)}$$

we observe that

$$d(z',z_o) = \left| \int_{v_\xi(z_o)}^{v_\xi(z)} \frac{dv}{\omega_\xi(v)} \right| ,$$

whence by (7.5)

$$d(z,z_o) \gg \int_{v_\xi(z_o)}^{v_\xi(z)} \frac{dv}{v} = \left| \log \frac{v_\xi(z)}{v_\xi(z_o)} \right|$$

as $v_\xi(z)$ and $v_\xi(z_o)$ tend to zero. If $D(z_o)$ denotes the hyperbolic disc with center z_o and radius 1 we thus obtain from (4.20)

$$V_\xi(z_o,s,\lambda) \ll \int_{D(z_o)} |V_\xi(z,s,\lambda)|\, d\omega(z) , \quad v_\xi(z_o) \to 0 , \tag{7.8}$$

where the implicit constant depends locally uniformly on s. Since Γ is discontinuous and $|V_\xi(z,s,\lambda)| \leqslant |V_\xi(z,\sigma,o)|$ by (4.2) and (4.8), the lemma therefore follows from (4.13), (7.7) and (7.8).

Now let $P_\xi^{1-\psi}(z,s,m)$ be defined by the right hand side of (7.2) with $1-\psi$ in place of ψ. The discreteness of Γ and (7.3) imply that $P_\xi^{1-\psi}$ is always given by a finite sum. Likewise only finite sums arise if in (7.2) ψ is replaced by one of its derivatives. We conclude that $P_\xi^\psi(\cdot,s,m)$ is a bounded C^∞-function on $\mathfrak{H}$ and

$$P_\xi(z,s,m) = P_\xi^\psi(z,s,m) + P_\xi^{1-\psi}(z,s,m) \tag{7.9}$$

if $\sigma > 1$. Therefore the series in (7.1) converge absolutely and locally uniformly for z in the complement of $\bigcup_{M\in\Gamma} MM_\xi^{-1}(\mathfrak{F}_\xi)$ and s

in $\sigma > 1$. Moreover $P_\xi(\cdot,s,m)$ are Γ-automorphic functions satisfying

$$\Delta P_\xi(z,s,m) = s(s-1)P_\xi(z,s,m)$$

for z and s in the just mentioned sets.

Remark 1. If κ is as in section 2 there are exactly κ cusps $\vartheta_1,\ldots,\vartheta_\kappa$ forming a complete set of representatives of all cusps with respect to the equivalence relation (2.7) (cf. [12]). Then the Eisenstein series $E_\iota(z,s)$ are by definition (cf. (4.16) and (7.1))

$$E_\iota(z,s) = \frac{1}{B(1/2,s)} P_{\vartheta_\iota}(z,s,o) \quad , \quad \iota = 1,\ldots,\kappa \ .$$

Remark 2. The results of section 9 will show that, up to a constant multiple,

$$(z,\zeta) \longmapsto P_\zeta(z,s,o)$$

is the resolvent kernel of Δ to the eigenvalue $s(s-1)$.

In order to give the Fourier series expansion of P_ξ with respect to χ we introduce in the notation of (2.4) - (2.7):

$$\delta(\xi,\chi) = \begin{cases} 1 \ , & \text{if } \ \xi \approx \chi \ \text{ and } \ \xi = \zeta \ \text{ or } \ \vartheta \ , \\ 2 \ , & \text{if } \ \xi = \eta \ \text{ and } \ \xi \approx \chi \ , \\ 0 \ , & \text{otherwise,} \end{cases}$$

$$\delta^*(\xi,\chi) = \begin{cases} 2 \ , & \text{if } \ \xi = \eta \ \text{ and } \ \xi^* \approx \chi \\ 0 \ , & \text{otherwise} \ . \end{cases} \tag{7.10}$$

Moreover let

$$_\xi p_\chi(s,m,n) = \frac{1}{\lambda_\chi} \sum_\nu {}_\xi s_\chi(m,n,\nu)\, {}_\xi i_\chi(\nu,s,\frac{m}{\lambda_\xi},\frac{n}{\lambda_\chi}) \tag{7.11}$$

and

$$_\xi P_\chi(s,m,n) = \frac{1}{\lambda_\chi} \sum_{\delta,\delta'=o}^{1} \varepsilon^\delta \varepsilon'^{\delta'} \sum_\nu {}_\xi^\delta S_\chi^{\delta'}(m,n,\nu)\, {}_\xi J_\chi(\nu,s,(-1)^\delta \frac{m}{\lambda_\xi},(-1)^{\delta'} \frac{n}{\lambda_\chi}) \ , \tag{7.12}$$

where the terms on the right hand sides are given by (5.10), (5.11), Lemma 7 and Lemma 11. By Lemma 6(ii) a finite sum stands in (7.11). The remark preceding Lemma 10 applies for the right hand side of (7.12) which will be shown to converge absolutely for s in $\sigma > 1$.

Proposition 1. Let $\mathfrak{U}_\chi$ be as in (6.3) with a positive ν_o such that ${}_\xi\nu_\chi(M) \geq \nu_o$ for every $\Gamma'_\xi M \Gamma'_\chi$ in ${}_\xi\Gamma_\chi$ and $\nu_o > 1$ unless $\xi = \vartheta$ or $\chi = \vartheta'$. With η^* as in (2.4) let ϱ_η be defined by $R_\vartheta M_\eta M_{\eta^*}^{-1} = \pm \exp(-\varrho_\eta X_\eta)$. If $\sigma > 1$ and z is in $\mathfrak{U}_\chi$ we have

$$S_\chi^{(1)} P_\xi(M_\chi^{-1}(z),s,m) = {}_\xi\mathcal{D}_\chi(z,s,m) + {}_\xi\mathcal{P}_\chi(z,s,m) + {}_\xi\mathfrak{P}_\chi(z,s,m) \quad ,$$

where

$${}_\xi\mathcal{D}_\chi(z,s,m) = \delta(\xi,\chi)V_\xi(z,s,\tfrac{m}{\lambda_\xi}) + \delta^*(\xi,\chi)\varepsilon e(\tfrac{m}{\lambda_\xi}\varrho_\xi)V_\xi(z,s,\tfrac{m}{\lambda_\xi}) \quad ,$$

$${}_\xi\mathcal{P}_\chi(z,s,m) = \sum_{M\in {}_\xi\gamma_\chi} \; \sum_{N\in\Gamma'_\chi/\mathbf{Z}_\Gamma} S_\chi^{(1)} V_\xi(z_{MN},s,\tfrac{m}{\lambda_\xi}) \tag{7.13}$$

and

$${}_\xi\mathfrak{P}_\chi(z,s,m) = \sum_{n\in Z} {}_\xi P_\chi(s,m,n)U_\chi(z,s,\tfrac{n}{\lambda_\chi}) \quad . \tag{7.14}$$

The sum in (7.13) converges absolutely and locally uniformly for z in the complement of $\bigcup_{N\in\Gamma'_\chi} \bigcup_{M\in {}_\xi\gamma_\chi} NM^{-1}(\widetilde{\mathfrak{F}}_\xi)$ and s in $\sigma > 0$. Moreover

$${}_\xi\mathcal{P}_\chi(z,s,m) - \gamma_\xi(s,\tfrac{m}{\lambda_\xi})\,{}_\xi\mathcal{P}_\chi(z,1-s,m) = \operatorname{ctg}\pi s \sum_{n\in Z} {}_\xi p_\chi(s,m,n)U_\chi(z,s,\tfrac{n}{\lambda_\chi}) \quad ,$$

where the sum on the right hand side converges absolutely and locally uniformly for z in $\mathfrak{H}$ and s in $0<\sigma<1$. Finally the sums in (7.12) and (7.14) converge absolutely and locally uniformly for z in $\mathfrak{U}_\chi$ and s in $\sigma > 1$.

Proof: Since by definition $M_\xi\Gamma_\xi = \Gamma'_\xi M_\xi$ we have

$$P_\xi(M_\chi^{-1}(z),s,m) = \sum_{M\in\Gamma'_\xi\backslash M_\xi\Gamma M_\chi^{-1}} V_\xi(z_M,s,\tfrac{m}{\lambda_\xi}) = \sum_{M\in\Gamma'_\xi\backslash M_\xi\Gamma M_\chi^{-1}/\Gamma'_\chi}$$

$$\sum_{N\in\Gamma'_\chi/\Gamma'_\chi\cap M^{-1}\Gamma'_\xi M} V_\xi(z_{MN},s,\tfrac{m}{\lambda_\xi}) \quad .$$

For $\Gamma'_\xi M = \Gamma'_\xi MN$ if and only if N belongs to $M^{-1}\Gamma'_\xi M$.

Suppose now that $\Gamma'_\chi \cap M^{-1}\Gamma'_\xi M$ is different from the center Z_Γ of Γ, i.e. suppose that $M_\chi(\chi)$ and $M^{-1}M_\xi(\xi)$ are fixed by a matrix N in $\Gamma'_\chi - Z_\Gamma$. By (3.4) and (5.1) this can happen only if $\mathfrak{f}_\chi = \mathfrak{f}_\xi$. This means that M leaves $\mathfrak{f}_\chi$ invariant and therefore represents a double coset of ${}_\xi\sigma_\chi$ defined in (5.9). Conversely, if $\Gamma'_\xi M \Gamma'_\chi$ is in ${}_\xi\sigma_\chi$, Lemma 6(iii) shows that $\Gamma'_\chi \cap M^{-1}\Gamma'_\xi M = \Gamma'_\chi$. For $\Gamma'_\xi = \Gamma'_\chi$ if ${}_\xi\sigma_\chi$ is non-empty. Thus we obtain from (5.7) - (5.9)

$$S^{(1)}_\chi P_\xi(M_\chi^{-1}(z),s,m) = {}_\xi\vartheta_\chi(z,s,m) + {}_\xi\mathfrak{P}_\chi(z,s,m) + {}_\xi\mathfrak{B}_\chi(z,s,m) ,$$

where ${}_\xi\mathfrak{P}_\chi$ is as in (7.13) and the other terms on the right hand side are defined by the series

$$\begin{aligned} {}_\xi\vartheta_\chi(z,s,m) &= \sum_{M\in {}_\xi\sigma_\chi} S^{(1)}_\chi V_\xi(z_M,s,\frac{m}{\lambda_\xi}) \\ {}_\xi\mathfrak{B}_\chi(z,s,m) &= \sum_{M\in {}_\xi\Gamma_\chi} \sum_{N\in\Gamma'_\chi/Z_\Gamma} S^{(1)}_\chi V_\xi(z_{MN},s,\frac{m}{\lambda_\xi}) \end{aligned} \tag{7.15}$$

in $\sigma > 1$.

We infer from (2.6), (2.7), (4.1), (4.13), Lemma 6(iii) and (7.10) that

$${}_\xi\vartheta_\chi(z,s,m) = \delta(\xi,\chi)V_\xi(z,s,\frac{m}{\lambda_\xi}) + \delta^*(\xi,\chi)V_\xi(M_\xi M_{\xi^*}^{-1}(z),s,\frac{m}{\lambda_\xi}) .$$

Since by (4.12), (4.14) and the definition of ϱ_η for $\eta = (\eta_1,\eta_2,\varepsilon)$ we have

$$V_\eta(M_\eta M_{\eta*}^{-1}(z),s,\lambda) = \varepsilon V_\eta(\exp(-\varrho_\eta X_\eta)(z),s,-\lambda) = \varepsilon e(\lambda\varrho_\eta)V_\eta(z,s,-\lambda)$$

our assertion on ${}_\xi\vartheta_\chi$ follows.

In dealing with ${}_\xi\mathfrak{P}_\chi$ we may assume that $(\xi,\chi) = (\eta,\zeta')$, (ζ,η') or (η,η') by Lemma 6(ii). In the first case (7.13) is given by a finite sum. Otherwise $\chi = \eta'$ and then (3.7), (4.13) and (4.20) yield

$$\sum_{N\in\Gamma'_\chi} V_\xi(\exp(\nu X_\zeta)N(z),s,\lambda) << \sum_{n\in Z}\left(\frac{ye^{n\lambda_\chi}}{\left|-ze^{n\lambda_\chi}\sin\frac{\nu}{2}+\cos\frac{\nu}{2}\right|^2}\right)^\sigma < \infty$$

if $0<|\nu-\pi|<\pi$, $\sigma>0$ and z in the complement of

$\bigcup_{N\in\Gamma'_\chi} N\exp(-\nu X_\zeta)(\mathfrak{F}_\xi)$. Since $V_\xi(\cdot,s,\lambda)$ is real-analytic on $\mathfrak{H}-\mathfrak{F}_\xi$ and since there are only finitely many double cosets in ${}_\xi\gamma_\chi$ by Lemma 6 (ii) the sum in (7.13) converges as asserted. By Lemma 4 we have

$$ {}_\xi\mathcal{P}_\chi(z,s,m) - \gamma_\xi(s,\tfrac{m}{\lambda_\xi})\,{}_\xi\mathcal{P}_\chi(z,1-s,m) = \operatorname{ctg}\pi s \sum_{M\in{}_\xi\gamma_\chi}\ \sum_{N\in\Gamma'_\chi/Z_\Gamma} S_\chi^{(1)} U_\xi(z_{MN},1-s,\tfrac{m}{\lambda_\xi}) \tag{7.16}$$

if $0<\sigma<1$. The inner sum on the right of (7.16) equals

$$\frac{1}{\lambda_\chi}\sum_{n\in Z}\int_{\mathcal{C}_\chi} e(-\tfrac{n}{\lambda_\chi}\varrho)\,S_\chi^{(1)}U_\xi\left(M\exp(\varrho X_\chi)(z),1-s,\tfrac{m}{\lambda_\xi}\right)d\varrho$$

by Fourier series inversion if $\chi=\zeta'$ and by Poisson's summation formula otherwise. The preceding sum converges absolutely and locally uniformly for z in $\mathfrak{H}$ and s in $0<\sigma<1$ since the inner sum on the right of (7.16) is real-analytic there. Now we insert the representation of M given after (5.11). By (4.11), a simple substitution and Lemma 11 the integral in the preceding sum equals

$$e(\tfrac{m}{\lambda_\xi}\,{}_\xi\Lambda_\chi^\ell(M)+\tfrac{n}{\lambda_\xi}\,{}_\xi\Lambda_\chi^r(M))\,{}_\xi i_\chi(\nu,s,\tfrac{m}{\lambda_\xi},\tfrac{n}{\lambda_\chi})\,U_\chi(z,s,\tfrac{n}{\lambda_\chi}) \ .$$

Thus the functional equation for ${}_\xi\mathcal{P}_\chi$ follows from (5.11), (7.11) and (7.16).

Finally, it follows from Lemma 1(i), (5.7), Lemma 12, (7.9) and the definition of $\mathfrak{U}_\chi$ that the sum in (7.15) converges absolutely and locally uniformly for z in $\mathfrak{U}_\chi$ and s in $\sigma>1$. Thus as before the inner sum on the right of (7.15) equals

$$\frac{1}{\lambda_\chi}\sum_{n\in Z}\int_{\mathcal{C}_\chi} e(-\tfrac{n}{\lambda_\chi}\varrho)\,S_\chi^{(1)}V_\xi(M\exp(\varrho X_\chi)(z),s,\tfrac{m}{\lambda_\xi})\,d\varrho \ .$$

By (5.7) M now belongs to ${}_\xi G_\chi$ and can be decomposed according to

Lemma 1(i). Together with (4.12) and (6.2) this lemma shows that the integral in the preceding sum equals

$$e\left(\frac{m}{\lambda_\xi}\,{}_\xi\Lambda^{\ell}_\chi(M)+\frac{n}{\lambda_\chi}\,{}_\xi\Lambda^{r}_\chi(M)\right)\int_{\mathscr{C}_\chi} e\left(-\frac{n}{\lambda_\chi}\varrho\right)S^{(1)}_\chi V_\xi\left(R^{\delta}_{\vartheta_\xi}N_\chi R^{-\delta'}_{\vartheta}\exp(\varrho X_\chi)(z),s,\frac{m}{\lambda_\xi}\right)d\varrho ,$$

where $\delta = {}_\xi\delta_\chi(M)$, $\delta' = {}_\xi\delta'_\chi(M)$ and in (6.2) $\nu = {}_\xi\nu_\chi(M)$. Since $\delta = 0$ unless $\xi=\eta$, $\delta'= 0$ unless $\chi = \eta'$ and $R_\vartheta \exp(\varrho X_\eta) = \exp(-\varrho X_\eta)R_\vartheta$ by (3.2), (3.16) the last integral is equal to

$$\varepsilon^{\delta}\varepsilon'^{\delta'}\,{}_\xi J_\chi\left(\nu,s,(-1)^{\delta}\frac{m}{\lambda_\xi},(-1)^{\delta'}\frac{n}{\lambda_\chi}\right)U_\chi\left(z,s,\frac{n}{\lambda_\chi}\right)$$

by (4.14) and Lemma 7. On summing the so obtained expression for the inner sum in (7.15) over ${}_\xi\Gamma_\chi$ we see from (5.10) and (7.12) that the right hand sides of (7.14) and (7.15) agree if z is in $\mathfrak{U}_\chi$ and s in $\sigma > 1$. There at least, the necessary interchange of summations is permissible. For the properties of convergence we stated above for (7.15) and simple facts on Fourier series readily imply our assertions on the convergence of the sums in (7.12) and (7.14). The proposition now follows.

Remark. By (4.16), (7.10), Remark 1 of this section and by Proposition 1 the Eisenstein series have Fourier series expansions of the form

$$S^{(1)}_\xi E_\iota(M^{-1}_\xi(z),s) = \delta(\xi,\vartheta_\iota)y^{s} + \sum_{n\in Z}\alpha_{\iota\xi}(s,n)U_\xi\left(z,s,\frac{n}{\lambda_\xi}\right) , \tag{7.17}$$

where

$$\alpha_{\iota\xi}(s,n) = \frac{{}_{\vartheta_\iota}P_\xi(s,o,n)}{B(1/2,s)} . \tag{7.18}$$

Moreover

$$\bar{\alpha}_{\iota\xi}(s,n) = \alpha_{\iota\xi}(1-s,-n) \tag{7.19}$$

on $\sigma = 1/2$ since $E_\iota(z,s)$ is positive on $s>1$ and $\bar{U}_\xi(z,s,\lambda) = U_\xi(z,\bar{s},-\lambda)$ by (4.7).

8. Computation of some integrals II

Let

$$\Psi_\xi(s,s',m) = \int_{\mathfrak{S}_\xi} V_\xi\left(z,s,\frac{m}{\lambda_\xi}\right)\bar{U}_\xi\left(z,s',\frac{m}{\lambda_\xi}\right)\psi_o(v_\xi(z))\,d\omega(z) , \tag{8.1}$$

where ψ_o is as in (7.3) and $\mathfrak{S}_\xi$ as in (7.6). By (4.20) and

(7.3)-(7.6) this integral converges absolutely for s in $\sigma > 1$ provided that

$$U_\xi(z,s',\frac{m}{\lambda_\xi}) \ll 1 \tag{8.2}$$

for $v_\xi(z) \to 0$. Now (4.6) yields

$$Y_\eta(\varrho,-\bar{z}) \ll (\mathrm{ch}\varrho)^{-2}$$

and, if $\xi = \zeta$ or η,

$$Y_\xi(\varrho,z) \ll (v_\xi + |\mathrm{sh}\frac{\varrho+u_\xi}{2}|)^{-2}$$

uniformly for ϱ on $\mathcal{b}_\xi$ and $v_\xi = v_\xi(z) \leq 1$. Thus we conclude from (4.1), (4.7), (4.15) and (4.23) that (8.2) holds for $s' = \sigma' + it'$ in $\frac{1}{2} \leq \sigma' \leq 1$ unless $\xi = \vartheta$, $m = 0$ and $s' = \frac{1}{2}$. Furthermore let

$$\psi_\xi(s,s',m) = \int_{\mathfrak{S}_\xi} \{V_\xi(z,s,\frac{m}{\lambda_\xi}) \frac{\partial}{\partial v_\xi} \bar{U}_\xi(z,s',\frac{m}{\lambda_\xi}) -$$

$$- \bar{U}_\xi(z,s',\frac{m}{\lambda_\xi}) \frac{\partial}{\partial v_\xi} V_\xi(z,s,\frac{m}{\lambda_\xi})\}\psi_o'(v_\xi)\,du_\xi dv_\xi , \tag{8.3}$$

where ψ_o' denotes the derivative of ψ_o. Since ψ_o' has support in the interval $[1,3/2]$ the results of section 4 show that $\psi_\xi(s,s',m)$ depends analytically on s in $\sigma > 1$ if s' lies in $\frac{1}{2} \leq \sigma' \leq 1$ and not simultaneously $\xi = \vartheta$, $m = 0$ and $s' = 1/2$. The following lemma relates Ψ_ξ and ψ_ξ.

Lemma 13. We have

$$\Psi_\xi(s,s',m) = \frac{\psi_\xi(s,s',m)}{(s-\bar{s}')(s-1+\bar{s}')}$$

and $\psi_\xi(s,s',m)$ is analytic in $\sigma > 0$ for $\frac{1}{2} \leq \sigma' \leq 1$ except when $\xi = \vartheta$, $m = 0$ and $s' = 1/2$. Moreover

$$\psi_\xi(1-\bar{s}',s',m) = \gamma_\xi(1-\bar{s}',\frac{m}{\lambda_\xi})\,\psi_\xi(\bar{s}',s',m) , \quad \psi_\xi(\bar{s}',s',m) = 2\pi\beta_\xi\lambda_\xi ,$$

where β_ξ is given by (4.19). In particular, if $\frac{1}{2} \leq \sigma' \leq 1$ and $s' \neq 1/2$, then $(2s-1)\,\Psi_\xi(s,s',m)$ is analytic in $\sigma > 0$ except for simple poles at $s = \bar{s}'$ with residue $2\pi\beta_\xi\lambda_\xi$ and at $s = 1-\bar{s}'$ with residue $2\pi\beta_\xi\lambda_\xi\ \gamma_\xi\left(1-\bar{s}', \frac{m}{\lambda_\xi}\right)$. On the other hand, $\Psi_\xi(s,\frac{1}{2},m)$ is analytic in $\sigma > 0$ except for a double pole at $s = 1/2$ with principal part

$$2\pi\beta_\xi\lambda_\xi\left\{\frac{1}{(s-1/2)^2} + \frac{1}{2s-1}\frac{\partial\gamma_\xi}{\partial s}\left(\frac{1}{2}, \frac{m}{\lambda_\xi}\right)\right\} ,$$

unless $\xi = \vartheta$ and $m = 0$, while $(2\bar{s}'-1)\Psi_\vartheta(s,s',0)$ is analytic in $\sigma > 0$ except for a simple pole at $s = \bar{s}'$ with residue 2π even for $s' = 1/2$.

<u>Proof</u>: It follows from (4.15), (4.16), (5.2), (7.6) and (8.1) that

$$\Psi_\vartheta(s,s',0) = B(1/2,s)\bar{B}(1/2,s'-1/2)\int_0^\infty y^{s-\bar{s}'-1}\psi_0(y)\,dy$$

$$= \frac{-2\pi}{(s-\bar{s}')(2\bar{s}'-1)}\frac{\Gamma(s)\,\Gamma(\bar{s}'+1/2)}{\Gamma(\bar{s}')\,\Gamma(s+1/2)}\int_0^\infty y^{s-\bar{s}'}\psi_0'(y)\,dy . \quad (8.4)$$

Since (8.4) implies the lemma for $\xi = \vartheta$, $m = 0$ we exclude this case in the following.

On integrating by parts we obtain from (8.1)

$$\bar{s}'(\bar{s}'-1)\Psi_\xi(s,s',m) = \int_{\mathfrak{F}_\xi}\bar{U}_\xi(z,s',\frac{m}{\lambda_\xi})\,\Delta\{V_\xi(z,s,\frac{m}{\lambda_\xi})\,\psi_0(v_\xi(z))\}\,d\omega(z)$$

$$= s(s-1)\Psi_\xi(s,s',m) + \int_{\mathfrak{F}_\xi}\bar{U}_\xi(z,s',\frac{m}{\lambda_\xi})\,\{2\psi_0'(v_\xi)\frac{\partial}{\partial v_\xi}V_\xi(z,s,\frac{m}{\lambda_\xi}) + \quad (8.5)$$

$$+\ \psi_0''(v_\xi)V_\xi(z,s,\frac{m}{\lambda_\xi})\}\,du_\xi\,dv_\xi .$$

In (8.5) we replace ψ_0'' by ψ_0' on integrating by parts with respect to v_ξ. In view of (8.3) this gives

$$(s-\bar{s}')(s-1+\bar{s}')\Psi_\xi(s,s',m) = \psi_\xi(s,s',m) .$$

The functional equation for V_ξ from Lemma 4 and (8.5) also yield

$$\psi_\xi(s,s',m) = \gamma_\xi(s,\frac{m}{\lambda_\xi})\,\psi_\xi(1-s,s',m) + \operatorname{ctg}\pi s\int_{\mathfrak{F}_\xi}\{U_\xi(z,1-s,\frac{m}{\lambda_\xi})\frac{\partial}{\partial v_\xi}\bar{U}_\xi(z,s',\frac{m}{\lambda_\xi})$$

$$-\ \bar{U}_\xi(z,s',\frac{m}{\lambda_\xi})\,\frac{\partial}{\partial v_\xi}U_\xi(z,1-s,\frac{m}{\lambda_\xi})\}\,\psi_0'(v_\xi)\,du_\xi\,dv_\xi . \quad (8.6)$$

By (4.7) and (4.11) the curly brackets in (8.6) vanish identically for $s = 1-\bar{s}'$. Thus we obtain

$$\psi_\xi(1-\bar{s}',s',m) = \gamma_\xi(1-\bar{s}', \frac{m}{\lambda_\xi})\,\psi_\xi(\bar{s}',s',m)$$

and

$$\frac{\partial\psi_\xi}{\partial s}(1/2,1/2,m) = \psi_\xi(1/2,1/2,m)\,\frac{\partial\gamma_\xi}{\partial s}(1/2, \frac{m}{\lambda_\xi}) - -\gamma_\xi(1/2, \frac{m}{\lambda_\xi})\frac{\partial\psi_\xi}{\partial s}(1/2,1/2,m).$$

Since $\gamma_\xi(1/2,\lambda) = 1$ unless $\xi = \vartheta$, $\lambda = 0$ we conclude that

$$\Psi_\xi(s,1/2,m) = \frac{\psi_\xi(s,1/2,m)}{(s-1/2)^2} = \psi_\xi(1/2,1/2,m)\left\{\frac{1}{(s-1/2)^2} + + \frac{1}{2s-1}\,\frac{\partial\gamma_\xi}{\partial s}(1/2, \frac{m}{\lambda_\xi})\right\} + O(1)$$

as $s \to 1/2$. Hence it remains to show that

$$\psi_\mathfrak{f}(\bar{s},s,m) = 2\pi\beta_\mathfrak{f}\lambda_\mathfrak{f} \ .$$

We deduce from the proof of Lemma 12 that in the various coordinate systems the Laplacian is given by

$$\Delta = 4y^2\,\frac{\partial^2}{\partial z\,\partial\bar{z}} = 4y^2\,\left|\frac{dz}{dw_\xi}\right|^{-2}\,\frac{\partial^2}{\partial w_\xi\,\partial\bar{w}_\xi} = \omega_\xi^2(v_\xi)\left(\frac{\partial^2}{\partial u_\xi^2} + \frac{\partial^2}{\partial v_\xi^2}\right). \qquad (8.7)$$

Since

$$\frac{\partial}{\partial u_\xi}V_\xi(z,s,\lambda) = 2\pi i\lambda\,V_\xi(z,s,\lambda)$$

and

$$\frac{\partial}{\partial u_\xi}\bar{U}_\xi(z,s,\lambda) = -2\pi i\lambda\bar{U}_\xi(z,s,\lambda)$$

by (4.11) and (4.12) the curly brackets in (8.3) do not depend on u_ξ while their derivative with respect to v_ξ equals

$$V_\xi(z,s,\lambda)\,\frac{\partial^2}{\partial v_\xi^2}\bar{U}_\xi(z,s',\lambda) - \bar{U}_\xi(z,s',\lambda)\,\frac{\partial^2}{\partial v_\xi^2}V_\xi(z,s,\lambda)$$

$$= \frac{1}{\omega_\xi^2(v_\xi)}\{V_\xi(z,s,\lambda)\,\Delta\bar{U}_\xi(z,s',\lambda) - \bar{U}_\xi(z,s',\lambda)\,\Delta V_\xi(z,s,\lambda)\} =$$

$$= \frac{s'(s'-1)-s(s-1)}{\omega_\xi^2(v_\xi)}\cdot V_\xi(z,s,\lambda)\bar{U}_\xi(z,s',\lambda)$$

with $\lambda = \frac{m}{\lambda_\xi}$. Thus the curly brackets in (8.3) are constant on $\mathfrak{H}$

if $s = \bar{s}'$. By (7.3) and (7.6) we have to show that the said constant is in fact equal to $-2\pi\beta_\xi$.

By elementary calculus we obtain from (4.6)

$$\frac{\partial}{\partial v_\xi} Y_\xi(\varrho,z) = Y_\xi(\varrho,z) f_\xi(v_\xi) - \frac{2}{\beta_\xi} Y_\xi^2(\varrho,z)$$

with

$$f_\xi(v) = \begin{cases} \mathrm{cthv}\ , & \text{if } \xi = \zeta\ , \\ 1/v\ , & \text{if } \xi = \vartheta\ , \\ \mathrm{ctgv}\ , & \text{if } \xi = \eta\ . \end{cases} \tag{8.8}$$

Consequently (4.7) and (4.8) yield

$$\frac{\partial}{\partial v_\xi} \bar{U}_\xi(z,s,\lambda) = \bar{s}\{\bar{U}_\xi(z,s,\lambda) f_\xi(v_\xi) - \frac{2}{\beta_\xi} \bar{U}_\xi(z,s+1,\lambda)\} \tag{8.9}$$

and

$$\frac{\partial}{\partial v_\xi} V_\xi(z,s,\lambda) = (1-s)\{V_\xi(z,s,\lambda) f_\xi(v_\xi) - \frac{2}{\beta_\xi} V_\xi(z,s-1,\lambda)\} \tag{8.10}$$

for $\xi = \zeta$ or ϑ . The last equation remains valid for $\xi = \eta$ as long as $v_\eta(z) < \frac{\pi}{2}$. However, since

$$\frac{\partial}{\partial v_\eta} TY_\eta(\varrho,z) = -T \frac{\partial}{\partial v_\eta} Y_\eta(\varrho,z) = Y_\eta(\varrho,-\bar{z}) f_\eta(v_\eta) + Y_\eta^2(\varrho,-\bar{z})$$

we have

$$\frac{\partial}{\partial v_\eta} \bar{U}_\eta(z,s,\lambda) = \bar{s}\{\bar{U}_\eta(z,s,\lambda)\,\mathrm{ctgv}_\eta - \bar{U}_{\tilde{\eta}}(z,s+1,\lambda)\}\ , \tag{8.11}$$

where $\tilde{\eta} = (\eta_1,\eta_2,-\varepsilon)$ if $\eta = (\eta_1,\eta_2,\varepsilon)$.

If $s = \bar{s}'$ let $C_\xi(s, \frac{m}{\lambda_\xi})$ denote the value of the curly brackets in (8.3) which does not depend on z as we already know. It follows from (8.9) and (8.10) that

$$\begin{aligned} C_\xi(s,\lambda) = {} & sV_\xi(z,s,\lambda)\{\bar{U}_\xi(z,\bar{s},\lambda) f_\xi(v_\xi) - \frac{2}{\beta_\xi} \bar{U}_\xi(z,\bar{s}+1,\lambda)\} \\ & + (s-1)\bar{U}_\xi(z,\bar{s},\lambda)\{V_\xi(z,s,\lambda) f_\xi(v_\xi) - \frac{2}{\beta_\xi} V_\xi(z,s-1,\lambda)\} \end{aligned}$$

unless $\xi = \eta$. Thus we deduce from (4.21), (4.22) and (8.8) that

$$C_\zeta(s,0) \sim 2\pi(s-1)(-2)\left\{\frac{\Gamma'}{\Gamma}(s) - \frac{\Gamma'}{\Gamma}(s-1)\right\} = -4\pi\ , \ v_\zeta \to \infty\ ,$$

while for $\lambda \neq 0$ (4.24), (4.25) and (8.8) yield

$$C_\zeta(s,\lambda) \sim \frac{s}{|\lambda|}\left\{1-\frac{B(2\pi|\lambda|,s)}{B(2\pi|\lambda|,s+1)}\right\} + \frac{s-1}{|\lambda|}\left\{1-\frac{B(2\pi|\lambda|,s-1)}{B(2\pi|\lambda|,s)}\right\} = -4\pi,$$

and $v_\zeta \to \infty$,

$$C_\vartheta(s,\lambda) \sim -s\Gamma(s)(\pi|\lambda|)^{-s}\frac{\pi}{\Gamma(s+1)}(\pi|\lambda|)^{s} - (s-1)(\pi|\lambda|)^{s-1}\Gamma(s-1)(\pi|\lambda|)^{1-s} =$$

$$= -2\pi\ , \ v_\vartheta \to \infty\ .$$

Finally, we infer from (4.26), (4.27), (8.8), (8.10) and (8.11) that for $\xi = (\eta_1,\eta_2,1)$

$$C_\xi(s,\lambda) \sim (1-s)\bar{U}_\xi(z,\bar{s},\lambda)V_\xi(z,s-1,\lambda) \sim -4\pi$$

and for $\xi = (\eta_1,\eta_2,-1)$

$$C_\xi(s,\lambda) \sim -sV_\xi(z,s,\lambda)\bar{U}_\xi(z,\bar{s}+1,\lambda) \sim -4\pi$$

as $v_\eta \uparrow \frac{\pi}{2}$. These asymptotic results show that in all cases $C_\xi(s,\lambda) = -2\pi\beta_\xi$.

Since $\psi_\xi(s,s',m)$ is analytic in $\sigma > 0$ by Lemma 4, (8.3) and (8.10) the lemma now follows.

<u>Lemma 14</u>. We have

$$\int_{\mathfrak{F}_\zeta} V_\zeta(z,s,0)\bar{U}_\zeta(z,s',0)\,d\omega(z) = \frac{4\pi\lambda_\zeta}{(s-\bar{s}')(s-1+\bar{s}')}\ ,$$

where the integral on the left converges absolutely in $\sigma > 1$ if $\frac{1}{2} \leq \sigma' \leq 1$.

<u>Proof</u>: The asserted convergence follows from (4.20)-(4.22), (7.4), (7.5) and (8.2). If $\mathfrak{F}(X)$ denotes the set of $u+iv$ in $\mathfrak{F}_\zeta$ with $v < X$, partial integrations, (7.4) and (8.7) yield

$$\int_{\mathfrak{F}(X)} \{\Delta V_\zeta(z,s,0)\bar{U}_\zeta(z,s',0) - V_\zeta(z,s,0)\Delta\bar{U}_\zeta(z,s',0)\}d\omega(z) = \int_0^{\lambda_\zeta}\{\bar{U}_\zeta(z,s',0)$$

$$\frac{\partial}{\partial v_\zeta}V_\zeta(z,s,0) - V_\zeta(z,s,0)\frac{\partial}{\partial v_\zeta}\bar{U}_\zeta(z,s',0)\}_{v_\zeta = X}\,du_\zeta\ . \qquad (8.12)$$

By (4.21), (4.22) and (8.8)-(8.10) the left hand side of (8.12) is

$$\sim \int_0^{\lambda_\xi} \bar{U}_\xi(z,s',o)(1-s)\{V_\xi(z,s,o) - V_\xi(z,s-1,o)\}du_\xi$$

$$\sim \int_0^{\lambda_\xi} 4\pi(1-s)\{\frac{\Gamma'}{\Gamma}(s-1) - \frac{\Gamma'}{\Gamma}(s)\}du_\xi = 4\pi\lambda_\xi$$

as $X \to \infty$. This clearly implies the lemma.

9. Analytic continuations and functional equations

In this section we show that all functions of s we met in section 7 are in fact meromorphic in the whole s-plane. Moreover they satisfy functional equations relating their values at s and $1-s$.

Proposition 2

(i) In the notation of Lemma 5, (7.2), (7.17) and (8.1) we have

$$P_\xi^\psi(z,s,m) = \sum_{j\geq o} \bar{\alpha}_{j\xi}(m)e_j(z)\Psi_\xi(s,s_j,m) + \frac{1}{4\pi i}\sum_{\iota=1}^{\kappa} \int_{(1/2)} \Big\{\bar{\alpha}_{\iota\xi}(s',m)\Psi_\xi(s,s',m) + \delta(\xi,\vartheta_\iota)\delta_{mo}\frac{\Psi_\xi(s,1-s',m)}{B(1/2,s'-1/2)}\Big\} E_\iota(z,s')ds' , \tag{9.1}$$

where δ_{mn} is the Kronecker symbol and the convergence is absolute and locally uniform for z in $\mathfrak{H}$ and s in $\sigma > 1$.

(ii) The right hand side of (9.1) actually converges absolutely and locally uniformly for z in $\mathfrak{H}$ and s in $\sigma > 1/2$. Moreover $P_\xi^\psi(z,s,m)$ is analytic in $\sigma \geq 1/2$ except for simple poles with residues

$$\frac{2\pi\beta_\xi\lambda_\xi}{2s-1}\sum_{s_\ell = s_j} \bar{\alpha}_{\ell\xi}(m)e_\ell(z)\cdot\begin{cases} 1 & \text{at } s=\bar{s}_j \neq 1/2 , \\ \gamma_\xi(1-\bar{s}_j, \frac{m}{\lambda_\xi}) & \text{at } s=1-\bar{s}_j \neq 1/2, \end{cases}$$

and except for a pole at $s = 1/2$ with principal part

$$\left\{\frac{1}{(s-1/2)^2} + \frac{1}{2s-1}\frac{\partial\gamma_\xi}{\partial s}(1/2,\frac{m}{\lambda_\xi})\right\} 2\pi\beta_\xi\lambda_\xi \sum_{s_j=1/2} \bar{\alpha}_{j\xi}(m)e_j(z) + \frac{\pi\beta_\xi\lambda_\xi}{2s-1}\sum_{\iota=1}^{\kappa} \bar{\alpha}_{\iota\xi}(\tfrac{1}{2},m)E_\iota(z,1/2) .$$

Here β_ξ , $\gamma_\xi(s,\lambda)$ and λ_ξ are as in (4.19), Lemma 4 and (5.2) respectively.

(iii) The functions $P_\xi(z,s,m)$ and $P^\psi_\xi(z,s,m)$ are meromorphic in s . Their difference is analytic for s in $\sigma > 0$. The Poincaré series satisfy the functional equations

$$P_\xi(z,1-s,m) = \gamma_\xi(1-s, \frac{m}{\lambda_\xi}) P_\xi(z,s,m) + \frac{2\pi\beta_\xi\lambda_\xi}{1-2s} \sum_{\iota=1}^{\kappa} \alpha_{\iota\xi}(1-s,-m) E_\iota(z,s) .$$

<u>Proof</u>: Let f be a Γ-invariant function such that for a given ξ

$$S^{(1)}_\xi f(M_\xi^{-1}(z)) = \beta\delta(\xi,\vartheta) y^{s'} + \sum_{n\in Z} \alpha_n U_\xi(z,s', \frac{n}{\lambda_\xi}) , \tag{9.2}$$

where the sum converges absolutely and locally uniformly on $\mathfrak{H}$. Since $\mathfrak{S}_\xi$ is a fundamental domain for Γ'_ξ we obtain from (7.2), Lemma 12, the G-invariance of ω and the Γ-invariance of f

$$<P^\psi_\xi(\cdot,s,m),f> = \sum_{M\in\Gamma'_\xi\backslash M_\xi\Gamma M_\xi^{-1}} \int_{M_\xi(\mathfrak{F})} V_\xi(z_M,s,\frac{m}{\lambda_\xi}) \psi(v_\xi(z_M)) \bar{f}(M_\xi^{-1}(z)) d\omega(z)$$

$$= \int_{\mathfrak{S}_\xi} V_\xi(z,s,\frac{m}{\lambda_\xi}) \psi(v_\xi(z)) \bar{f}(M_\xi^{-1}(z)) d\omega(z)$$

if $\sigma > 1$. By (4.1) and (5.4)

$$2f(M_\eta^{-1}(z)) = S^{(1)}_\eta f(M_\eta^{-1}(z)) + S^{(1)}_{\tilde\eta} f(M_{\tilde\eta}^{-1}(z))$$

if η and $\tilde\eta$ are as in (8.11). Hence (4.11), (4.12), (4.15), (7.3), (8.1) and (9.2) yield

$$\begin{aligned} <P^\psi_\xi(\cdot,s,m),f> &= \bar\alpha_m \Psi_\xi(s,s',m) + \beta\delta(\xi,\vartheta)\delta_{mo} \int_{\mathfrak{S}_\xi} V_\xi(z,s,\frac{m}{\lambda_\xi}) \overline{y^{s'}} \psi_o(v_\xi(z)) d\omega(z) \\ &= \bar\alpha_m \Psi_\xi(s,s',m) + \beta\delta(\xi,\vartheta)\delta_{mo} \frac{\Psi_\xi(s,1-s',m)}{B(1/2,1/2-\bar{s}')} . \end{aligned} \tag{9.3}$$

Since $P^{\psi}_{\xi}(\cdot,s,m)$ is a bounded C^{∞}-function on $\mathfrak{H}$ as we noted after Lemma 12, part (i) follows from (2.12), Lemma 5, (7.17) and (9.3).

By Cauchy's inequality and Lemma 13 the square of (9.1) is

$$<<\Big(\sum_{j\geq 0}|\alpha_{j\xi}(m)\psi_{\xi}(s,s_j,m)|^2+\frac{1}{4\pi i}\sum_{\iota=1}^{\kappa}\int_{(1/2)}\Big|\bar{\alpha}_{\iota\xi}(s',m)\psi_{\xi}(s,s',m)+\delta(\xi,\vartheta_{\iota})\delta_{mo}\frac{\psi_{\xi}(s,1-s',m)}{B(1/2,s'-1/2)}\Big|^2 ds'\Big)\Big(\sum_{j\geq 0}\frac{|e_j(z)|^2}{|(s-\bar{s}_j)(s-1+\bar{s}_j)|^2}+ \tag{9.4}$$

$$+\frac{1}{4\pi i}\sum_{\iota=1}^{\kappa}\int_{(1/2)}\frac{|E_{\iota}(z,s')|^2}{|(s-\bar{s}')(s-1+\bar{s}')|^2}ds'\Big).$$

Now let

$$W_{\xi}(z,s,\lambda) = \{2\psi'(v_{\xi}(z))\frac{\partial}{\partial v_{\xi}}V_{\xi}(z,s,\lambda) + \psi''(v_{\xi}(z))V_{\xi}(z,s,\lambda)\}\omega^2_{\xi}(v_{\xi}(z))$$

and

$$\tilde{P}_{\xi}(z,s,m) = \sum_{M\in\Gamma_{\xi}\backslash\Gamma} W_{\xi}(z_{M_{\xi}M},s,\frac{m}{\lambda_{\xi}}).$$

Since Γ is discrete and the derivatives of ψ_o have support in the interval $[1,3/2]$ $\tilde{P}_{\xi}$ is always given by a finite sum. Moreover by Lemma 4 and (8.10) $\tilde{P}_{\xi}(z,s,m)$ is analytic in $\sigma > 0$ and satisfies

$$\tilde{P}_{\xi}(z,s,m) << |s| \tag{9.5}$$

uniformly for z in $\mathfrak{H}$ and s in $\sigma_o \leq \sigma \leq 2$ if $\sigma_o > 0$. Similarly as for (9.3) we obtain from (7.4), (8.5) and (9.2)

$$<\tilde{P}_{\xi}(\cdot,s,m),f> = \int_{\mathfrak{S}_{\xi}} W_{\xi}(z,s,\frac{m}{\lambda_{\xi}})\bar{f}(M_{\xi}^{-1}(z))d\omega(z) = \bar{\alpha}_m\psi(s,s',m) + \beta\delta(\xi,\vartheta)\delta_{mo}\frac{\psi_{\xi}(s,1-s',m)}{B(1/2,1/2-s')}.$$

Thus the first factor on the right hand side of (9.4) is finite in $\sigma > 0$. For it equals

$$\int_{\mathfrak{F}}|\tilde{P}_{\xi}(z,s,m)|^2 d\omega(z)$$

by (2.13), Lemma 5 and (7.17). Similarly the second factor on the right of (9.4) is equal to

$$(2\lambda_{\zeta})^{-2}\int_{\mathfrak{F}}|P_{\zeta}(z',s,o)|^2 d\omega(z') \tag{9.6}$$

with $\zeta = z$. For by (7.1) and Lemma 14

$$<P_\zeta(\cdot,s,o),f> = \frac{4\pi\lambda_\zeta\alpha_o}{(s-\bar{s}')(s+\bar{s}'-1)}$$

if $\xi = \zeta$ in (9.2), while (3.7), (4.21), (4.24) and (5.1) yield

$$f(\zeta) = f(M_\zeta^{-1}(i)) = \alpha_o U_\zeta(i,s',o) = 2\pi\alpha_o .$$

In $\sigma > 1$ (9.6) is finite by Proposition 1, (4.15), (4.22) and (4.24). Thus the convergence of (9.1) is as asserted in part (ii) since

$$|(s-\bar{s}')(s+\bar{s}'-1)| >> (s+1-\bar{s}')(s+\bar{s}') \quad , \quad \mathrm{Im}\, s' \to \infty$$

for s in $0 \leqslant \sigma \leqslant 1$. We also note that the first sum on the right of (9.1) is analytic in $\sigma > 0$ except for poles due to the singularities of the individual summands.

Let $\mathcal{C}_-$ denote a simple path from $\frac{1}{2} - i\infty$ to $\frac{1}{2} + i\infty$ in the strip $\sigma^* < \sigma < 1/2$, where $\sigma^* = \max_{s_j > 1/2}(1-s_j)$, such that the singularities of $E_\iota(z,s)$, $\iota = 1,\ldots,\kappa$, in $\sigma < 1/2$ actually lie to the left of $\mathcal{C}_-$. Assume, moreover, that under complex conjugation $\mathcal{C}_-$ goes into $-\mathcal{C}_-$ and denote the image of $\mathcal{C}_-$ under $s \longmapsto 1-\bar{s}$ by $\mathcal{C}_+$. Thus, if s lies in $\sigma > 1/2$ and to the left of $\mathcal{C}_+$, we have by Cauchy's theorem, Lemma 4 and Lemma 13

$$\frac{1}{4\pi i}\left(\int_{(1/2)} - \int_{\mathcal{C}_+}\right)\left\{\alpha_{\iota\xi}(s',m)\bar{\Psi}_\xi(s,s',m) + \right.$$

$$\left. + \delta(\xi,\vartheta_\iota)\delta_{mo}\frac{\bar{\Psi}_\xi(s,1-s',m)}{B(1/2,1/2-s')}\right\}E_\iota(z,1-s')ds'$$

$$= \left\{\frac{\alpha_{\iota\xi}(\bar{s},m)\bar{\Psi}_\xi(s,\bar{s},m)}{2(2\bar{s}-1)} + \delta(\xi,\vartheta_\iota)\delta_{mo}\frac{\bar{\Psi}_\xi(s,1-\bar{s},m)}{2(2\bar{s}-1)}\right\}E_\iota(z,s) =$$

$$= \frac{\pi\beta_\xi\lambda_\xi}{2\bar{s}-1}\alpha_{\iota\xi}(\bar{s},m)E_\iota(z,1-\bar{s}) .$$

By (7.19) the contribution to (9.1) involving Eisenstein series therefore equals

$$\frac{\pi\beta_\xi\lambda_\xi}{2s-1}\sum_{\iota=1}^{\kappa}\alpha_{\iota\xi}(s,-m)E_\iota(z,1-s)+\frac{1}{4\pi i}\sum_{\iota=1}^{\kappa}\int_{\mathscr{C}_+}\Big\{\alpha_{\iota\xi}(s',-m)\Psi_\xi(s,\bar{s}',m)+ \tag{9.7}$$
$$+\delta(\xi,\vartheta_\iota)\delta_{mo}\frac{\Psi_\xi(s,1-\bar{s}',m)}{B(1/2,1/2-s')}\Big\}E_\iota(z,1-s')ds'$$

if s lies in $\sigma > 1/2$ and to the left of $\mathscr{C}_+$. For $\mathscr{C}_+$ goes into $-\mathscr{C}_+$ under complex conjugation. Lemma 13 and the definition of $\mathscr{C}_\pm$ now reveal that (9.7) is analytic for s in the region between $\mathscr{C}_-$ and $\mathscr{C}_+$ except possibly for a simple pole at $s = 1/2$. Hence the singularities of $P_\xi^\psi(z,s,m)$ in $\sigma \geqslant 1/2$ and their principal parts can be read off from (9.1) and (9.7) by means of Lemma 13. Note that our expression for the principal part remains valid in the 'exceptional' case $\xi = \vartheta$, $s = 1/2$ and $m = 0$ since $\alpha_{j\vartheta}(0) = 0$ for $s_j = 1/2$ by Lemma 5.

Our observations following (4.8) and Lemma 12 show that

$$P_\xi(z,s,m) - P_\xi^\psi(z,s,m) = P_\xi^{1-\psi}(z,s,m)$$

is analytic for s in $\sigma > 0$. By Proposition 1 we have

$$\sum_{n\in Z^\xi} {}_\xi P_\chi(s,m,n)U_\chi(z,s,\frac{n}{\lambda_\chi}) = S_\chi^{(1)}P_\xi^\psi(M_\chi^{-1}(z),s,m) + \{S_\chi^{(1)}P_\xi^{1-\psi}(M_\chi^{-1}(z),s,m) \tag{9.8}$$
$$-{}_\xi\mathscr{V}_\chi(z,s,m) - {}_\xi\mathscr{P}_\chi(z,s,m)\}$$

for z in $\mathfrak{U}_\chi$ and s in $\sigma > 1$. The left hand side and the first term on the right of (9.8) are C^∞-functions on $\mathfrak{U}_\chi$ if $\sigma > 1$. Thus by analytic continuation it follows from Proposition 1 that the curly brackets in (9.8) are C^∞ on $\mathfrak{U}_\chi$ if only s lies in $\sigma > 0$. Hence the preceding proof of Proposition 2(ii) also establishes the analytic continuation of ${}_\xi P_\chi(s,m,n)$ into the region to the right of $\mathscr{C}_-$. In particular, the left hand side of (9.8) converges absolutely and locally uniformly for z in $\mathfrak{U}_\chi$ and s in the region between $\mathscr{C}_-$ and $\mathscr{C}_+$.

Assume now that s lies in $\sigma > 1/2$ and to the left of $\mathcal{C}_+$. On using Proposition 1 for every $\chi = \xi'$ we conclude from (4.9) and Lemma 4 that

$$P_\xi(z,1-s,m) - \gamma_\xi(1-s,\frac{m}{\lambda_\xi})P_\xi(z,s,m) - B(1/2,1/2-s)\sum_{\iota=1}^{\kappa} {}_\xi P_{\vartheta_\iota}(1-s,m,o)E_\iota(z,s) \tag{9.9}$$

is an eigenfunction of Δ on all of $\mathfrak{H}$. Moreover we infer from (4.1), (4.15), (4.24), Lemma 4, (7.17) and Proposition 1 with $\chi = \vartheta'$ that, after replacing z by $M_\chi^{-1}(z)$, (9.9) equals

$$-\delta(\xi,\chi)\,\mathrm{ctg}\,\pi s\, U_\xi(z,s,\frac{m}{\lambda_\xi}) + {}_\xi P_\chi(1-s,m,o)U_\chi(z,1-s,o) -$$

$$-\gamma_\xi(1-s,\frac{m}{\lambda_\xi})\,{}_\xi P_\chi(s,m,o)U_\chi(z,s,o) - B(1/2,1/2-s)\sum_{\iota=1}^{\kappa} {}_\xi P_{\vartheta_\iota}(1-s,m,o)\{\delta(\vartheta_\iota,\chi)y^s$$

$$+ \alpha_{\iota\chi}(s,o)U_\chi(z,s,o)\} + o(1)$$

$$= {}_\xi P_\chi(1-s,m,o)\{U_\chi(z,1-s,o) - B(1/2,1/2-s)y^s\} + O(y^{1-\sigma}) = O(y^{1-\sigma})$$

as $y = \mathrm{Im}\,z$ tends to ∞. This shows that (9.9) belongs to $L_2(\Gamma\backslash\mathfrak{H})$, whence it has to vanish identically. For its eigenvalue $s(s-1)$ is non-real for the non-real s under consideration whereas the Laplacian is self-adjoint in $L_2(\Gamma\backslash\mathfrak{H})$.

The vanishing of (9.9) and (7.18) yield the functional equation for P_ξ if

$${}_\xi P_{\vartheta_\iota}(s,m,o) = \beta_\xi\lambda_\xi\,{}_{\vartheta_\iota}P_\xi(s,o,-m)\,,\quad \iota = 1,\dots,\kappa\,. \tag{9.10}$$

Since $\beta_\vartheta = \lambda_\vartheta = 1$ we deduce (9.10) from Lemma 8 and (7.12) provided that

$${}^{\delta}_{\xi}S^{\delta'}_{\chi}(m,n,\nu) = {}^{\delta'}_{\chi}S^{\delta}_{\xi}(-n,-m,\nu)\,.$$

This latter identity, however, follows from (3.15), Lemma 1 and (5.10) since

$${}_\xi\delta_\chi(M) = {}_\chi\delta'_\xi(M^{-1}) \text{ and } {}_\xi\nu_\chi(M) = {}_\chi\nu_\xi(M^{-1}) \quad \text{for every } M \text{ in } {}_\xi G_\chi\,.$$

We conclude from part (ii) and the functional equations just obtained that, together with the Eisenstein series, all $P_\xi(z,s,m)$

are meromorphic in s . This completes the proof of Proposition 2.

Remark 1. If $\xi = \vartheta$, $m = 0$ then $\gamma_\xi(s, \frac{m}{\lambda_\xi}) = 0$ and Proposition 2 (iii) gives the functional equations for the Eisenstein series. Moreover, (9.10) then expresses the symmetry of the so-called constant term matrix (cf. [12], Thm. 2.2.1). It should be noted, on the other hand, that Proposition 2 does not provide yet another proof for the analytic continuation of the Eisenstein series. For our derivation relies on this fact.

Remark 2. The idea of proving functional equations in the above manner is taken from [15]. We must point out, however, that Neunhöffer's functional equation for the Poincaré series to an inner point (Satz 6.4) is completely wrong as a result of a mistake he made in relating different hypergeometric functions.

Proposition 3

(i) The Fourier coefficient ${}_\xi p_\chi(s,m,n)$ is meromorphic in s . It is analytic for s in $0 < \sigma < 1$ and satisfies

$$\gamma_\chi(s, \frac{n}{\lambda_\chi})\, {}_\xi p_\chi(1-s,m,n) = \gamma_\xi(1-s, \frac{m}{\lambda_\xi})\, {}_\xi p_\chi(s,m,n) .$$

(ii) The Fourier coefficient ${}_\xi P_\chi(s,m,n)$ is analytic for s in $\sigma \geqslant 1/2$ except for simple poles with residues

$$\frac{2\pi\beta_\xi\lambda_\xi}{2s-1} \sum_{s_\ell = s_j} \bar{\alpha}_{\ell\xi}(m)\,\alpha_{\ell\chi}(n) \cdot \begin{cases} 1 & \text{at } s = s_j \text{ , if } \frac{1}{2} < s_j \leqslant 1 , \\ \gamma_\xi(s_j, \frac{m}{\lambda_\xi}) & \text{at } s = s_j = \frac{1}{2} + it_j \text{ , } t_j > 0 , \\ \gamma_\chi(1-s_j, \frac{n}{\lambda_\chi}) & \text{at } s = 1-s_j = \frac{1}{2} - it_j \text{ , } t_j > 0 , \end{cases}$$

and except for a pole at $s = 1/2$ with principal part

$$\left\{\frac{1}{(s-1/2)^2} + \frac{1}{2s-1}\left[\frac{\partial\gamma_\xi}{\partial s}\left(\frac{1}{2}, \frac{m}{\lambda_\xi}\right) + \frac{\partial\gamma_\chi}{\partial s}\left(1/2, \frac{n}{\lambda_\chi}\right)\right]\right\} 2\pi\beta_\xi\lambda_\xi \sum_{s_j = 1/2} \bar{\alpha}_{j\xi}(m)\,\alpha_{j\chi}(n) +$$

$$+ \frac{\pi\beta_\xi\lambda_\xi}{2s-1} \sum_{\iota=1}^{\kappa} \bar{\alpha}_{\iota\xi}(1/2,m)\,\alpha_{\iota\chi}(1/2,n) .$$

(iii) Every $_{\xi}P_{\chi}(s,m,n)$ is meromorphic in s and satisfies

$$\gamma_{\chi}(s,\frac{n}{\lambda_{\chi}})\{_{\xi}P_{\chi}(1-s,m,n)+\frac{1}{2}\operatorname{ctg}\pi(1-s)\,_{\xi}p_{\chi}(1-s,m,n)\} = \gamma_{\xi}(1-s,\frac{m}{\lambda_{\xi}})\{\,_{\xi}P_{\chi}(s,m,n)$$

$$+\frac{1}{2}\operatorname{ctg}\pi s\,_{\xi}p_{\chi}(s,m,n)\} + \operatorname{ctg}\pi s[\delta(\xi,\chi)\delta_{mn}+\delta^*(\xi,\chi)\delta_{-mn}\,\varepsilon\, e(\frac{m}{\lambda_{\xi}}\varrho_{\xi})]$$

$$+\frac{2\pi\beta_{\xi}\lambda_{\xi}}{1-2s}\sum_{\iota=1}^{\kappa}\alpha_{\iota\xi}(1-s,-m)\alpha_{\iota\chi}(s,n) \quad ,$$

where $\delta(\xi,\chi)$, $\delta^*(\xi,\chi)$ and ϱ_{ξ} are given in (7.10) and Proposition 1 respectively.

(iv) The Fourier coefficients of the Eisenstein series satisfy

$$\sum_{\iota=1}^{\kappa}\alpha_{\iota\xi}(1-s,-m)\alpha_{\iota\chi}(s,n) = \begin{cases} \dfrac{\gamma_{\chi}(s,\frac{n}{\lambda_{\chi}})}{\gamma_{\xi}(s,\frac{m}{\lambda_{\xi}})}\displaystyle\sum_{\iota=1}^{\kappa}\alpha_{\iota\xi}(s,-m)\alpha_{\iota\chi}(1-s,n)\,, \text{ unless} \\ \qquad \xi=\vartheta\,,\ m=0 \text{ or } \chi=\vartheta'\,,\ n=0\,, \\ \dfrac{s-1/2}{\pi}\{\delta(\xi,\chi)\delta_{mn}\operatorname{ctg}\pi s-\gamma_{\chi}(s,\frac{n}{\lambda_{\chi}})B(1/2,1-s) \\ \displaystyle\sum_{\iota=1}^{\kappa}\delta(\xi,\vartheta_{\iota})\alpha_{\iota\chi}(1-s,n)\}\,, \text{ if } \xi=\vartheta\,,\ m=0\,. \end{cases}$$

<u>Proof</u>: Part (i) immediately follows from Lemma 11 since (7.11) is a finite sum. We have already noted in the proof of Proposition 2 that, except for some poles, $_{\xi}P_{\chi}(s,m,n)$ is analytic to the right of $\mathscr{G}_-$. More precisely (4.11), Proposition 1 and (9.8) show that

$$_{\xi}P_{\chi}(s,m,n)U_{\chi}(z,s,\frac{n}{\lambda_{\chi}}) - \frac{1}{\lambda_{\chi}}\int_0^{\lambda_{\chi}} S_{\chi}^{(1)}P_{\xi}^{\psi}(M_{\chi}^{-1}\exp(\varrho X_{\chi})(z),s,m)e(-\frac{n}{\lambda_{\chi}}\varrho)d\varrho$$

is analytic for s in $\sigma > 0$ and z in $\mathfrak{U}_{\chi}$. Therefore Proposition 2 (ii), Lemma 5 and (7.17) allow to determine location and principal parts of the poles of $_{\xi}P_{\chi}(s,m,n)$ in $\sigma \geq 1/2$. In particular, $_{\xi}P_{\chi}(s,m,n)U_{\chi}(z,s,\frac{n}{\lambda\chi})$ has simple poles with residues

$$\frac{2\pi\beta_\xi\lambda_\xi}{2s-1}\sum_{s_\ell=s_j}\bar{\alpha}_{\ell\xi}(m)\alpha_{\ell\chi}(n)U_\chi(z,s_j,\frac{n}{\lambda_\chi})\cdot\begin{cases}1 & \text{at } s=\bar{s}_j\neq 1/2\,,\\ \gamma_\xi(1-\bar{s}_j,\frac{m}{\lambda_\xi}) & \text{at } s=1-\bar{s}_j\neq 1/2\,,\end{cases}$$

while at $s = 1/2$ it has a pole with principal part

$$\left\{\frac{1}{(s-1/2)^2}+\frac{1}{2s-1}\frac{\partial\gamma_\xi}{\partial s}(1/2,\frac{m}{\lambda_\xi})\right\}2\pi\beta_\xi\lambda_\xi\sum_{s_j=1/2}\bar{\alpha}_{j\xi}(m)\alpha_{j\chi}(n)U_\chi(z,1/2,\frac{n}{\lambda_\chi})$$

$$+\frac{\pi\beta_\xi\lambda_\xi}{2s-1}\sum_{\iota=1}^{\kappa}\bar{\alpha}_{\iota\xi}(1/2,m)\left\{\delta(\chi,\vartheta_\iota)y^{1/2}+\lim_{s\to 1/2}\alpha_{\iota\chi}(s,n)U_\chi(z,s,\frac{n}{\lambda_\chi})\right\}.$$

Hence part (ii) follows from Lemma 4. For $U_\chi(z,s,\lambda')$ is analytic in $\sigma \geqslant 1/2$, $s \neq 1/2$ and

$$\frac{\partial}{\partial s}U_\chi(z,s,\lambda')_{s=1/2} = U_\chi(z,1/2,\lambda')\frac{\partial\gamma_\chi}{\partial s}(1/2,\lambda')$$

unless $\chi = \vartheta'$, $\lambda' = 0$. In addition we note that ${}_\xi P_{\vartheta'}(s,m,o)$ is analytic at $s = 1/2$ since by Lemma 5 $\alpha_{j\vartheta'}(o) = 0$ for $s_j = 1/2$ and $\alpha_{\iota\vartheta'}(1/2,o) = 0$ for $\iota = 1,\dots,\kappa$. The latter equation holds because $E_\iota(z,s)$ is analytic and $U_\vartheta(z,s,o)$ has a pole at $s = 1/2$.

By (4.11), Proposition 1 and 2 we have

$${}_\xi P_\chi(s,m,n)U_\chi(z,s,\frac{n}{\lambda_\chi}) = \frac{1}{\lambda_\chi}\int_0^{\lambda_\chi}\{S_\chi^{(1)}P_\xi(M_\chi^{-1}\exp(\varrho X_\chi)(z),s,m)$$

$$-{}_\xi\mathcal{V}_\chi(\exp(\varrho X_\chi)(z),s,m) - {}_\xi\mathcal{P}_\chi(\exp(\varrho X_\chi)(z),s,m)\}e(-\frac{n}{\lambda_\chi}\varrho)d\varrho$$

for z in $\mathfrak{U}_\chi$ and s in the region between $\mathcal{L}_-$ and $\mathcal{L}_+$. On replacing s by $1-s$ and transforming the first term in the above integral by the functional equation of Proposition 2(iii) we obtain from Lemma 4, Proposition 1 and (7.17)

$${}_\xi P_\chi(1-s,m,n)U_\chi(z,1-s,\frac{n}{\lambda_\chi}) = \gamma_\xi(1-s,\frac{m}{\lambda_\xi})\,{}_\xi P_\chi(s,m,n)U_\chi(z,s,\frac{n}{\lambda_\chi})$$

$$-\operatorname{ctg}\pi(1-s)\,{}_\xi p_\chi(1-s,m,n)U_\chi(z,1-s,\frac{n}{\lambda_\chi}) + \operatorname{ctg}\pi s[\delta(\xi,\chi)\delta_{mn}U_\xi(z,s,\frac{m}{\lambda_\xi}) + \tag{9.11}$$

$$+\delta^*(\xi,\chi)\delta_{-mn}\varepsilon e(\frac{m}{\lambda_\xi}\varrho_\xi)U_\xi(z,s,-\frac{m}{\lambda_\xi})] + \frac{2\pi\beta_\xi\lambda_\xi}{1-2s}\sum_{\iota=1}^{\kappa}\alpha_{\iota\xi}(1-s,-m)(\delta(\chi,\vartheta_\iota)\delta_{no}y^s +$$

$$+\alpha_{\iota\chi}(s,n)U_\chi(z,s,\frac{n}{\lambda_\chi}))\,.$$

Since by (4.15) and Lemma 4

$$U_\chi(z,1-s,\lambda) = \gamma_\chi(s,\lambda)U_\chi(z,s,\lambda) + y^s B(1/2,1/2-s)\delta_{o\lambda}\sum_{\iota=1}^{\kappa}\delta(\chi,\vartheta_\iota)$$

we obtain part (iii) from part (i) by comparing the coefficients of $U_\chi(z,s,\frac{n}{\lambda_\chi})$ on both sides of (9.11). Note that, if $\chi = \vartheta'$, $n = 0$, the second identity arising from the coefficients of y^s is equivalent to (9.10).

Since by Lemma 4 $\gamma_\xi(s,\lambda)\gamma_\xi(1-s,\lambda) = 1$ unless $\xi = \vartheta$ and $\lambda = 0$, the first equation in part (iv) results from comparing (iii) after the substitution $s \longmapsto 1-s$ with (iii) itself. The second equation in (iv) directly follows from (7.18) and part (iii) since then $\delta^*(\xi,\chi) = {}_\xi p_\chi(s,m,n) = 0$. Thus Proposition 3 is established.

<u>Remark</u>. As a special case of Proposition 3(iv) we have

$$\sum_{\iota=1}^{\kappa} \alpha_{\iota\vartheta}(1-s,o)\alpha_{\iota\vartheta'}(s,o) = \frac{s-1/2}{\pi}\mathrm{ctg}\pi s\ \delta(\vartheta,\vartheta') .$$

By (4.15) and (7.17) this identity is equivalent to the functional equation for the so-called constant term matrix (cf. [12] Thm. 4.4.1).

A final lemma is needed to justify the shifting of contours later on.

<u>Lemma 15</u>. For every ξ,χ,m and n there exist a $C > o$ and a sequence $(\tau_\ell)_{\ell=1}^{\infty}$ with $\tau_\ell \to \infty$ such that

$$\int_{1/2}^{3/2} |{}_\xi P_\chi(\sigma+it,m,n)|\,d\sigma \ll |t|^C$$

for $t = \pm\tau_\ell$, $\ell = 1,2,\ldots$.

<u>Proof</u>: It is readily verified that the partial derivative of (4.6) with respect to ϱ vanishes if and only if $\varrho \equiv -u_\xi \pmod{\frac{\ell_\xi}{2}}$, where ℓ_ξ is as in (3.8). Moreover we observe that

$$Y_\xi(-u_\xi+i\tau,z) = Y_\xi(-u_\xi,z)\left(1 + \frac{\tau^2}{4\omega_\xi^2(v_\xi/2)} + O(\tau^3)\right) , \tau\to 0 ,$$

in the notation of (7.5). The method of the stationary phase says that the major contribution to an oscillatory integral comes from a neighbourhood of the stationary points and, in particular,

$$\int_{-\infty}^{\infty} e^{i\pi t\tau^2} \mathscr{W}(\tau)\,d\tau \sim e\left(\frac{t}{8|t|}\right) \frac{\mathscr{W}(o)}{|t|^{1/2}} , \quad |t| \to \infty ,$$

for every C^∞-function $\mathscr{W}$ of compact support on $\mathbb{R}$. Thus we obtain from (4.2), (4.6) and (4.8)

$$V_\xi(z,s,\lambda) = e(\lambda u_\xi) Y_\xi^{1-s}(-u_\xi,z) \int_{-v_\xi}^{v_\xi} e(-\lambda i\tau) \left(\frac{Y_\xi(-u_\xi+i\tau,z)}{Y_\xi(-u_\xi,z)} \right)^{1-s} d\tau$$

$$\sim e\left(\lambda u_\xi - \frac{t}{8|t|}\right) Y_\xi^{1-s}(-u_\xi,z) \frac{2\sqrt{\pi}}{|t|^{1/2}} \omega_\xi(v_\xi/2) , \quad |t| \to \infty ,$$

uniformly for s in $\frac{1}{2} \leqslant \sigma \leqslant 2$ provided that $v_\xi < \pi/2$ for $\xi = \eta$.
By Lemma 4 and Stirling's formula we have

$$\gamma_\xi(1-s,\lambda) \sim \begin{cases} (-1)^{2\pi\lambda} , & \text{if } \xi = \zeta , \\ \left(\frac{\pi|\lambda|}{|t|}\right)^{2s-1} e^{2it} , & \text{if } \xi = \vartheta \text{ and } \lambda \neq 0 , \\ e\left(-\frac{\varepsilon t}{4|t|}\right) , & \text{if } \xi = (\eta_1,\eta_2,\varepsilon) , \end{cases}$$

as $|t| \to \infty$, uniformly in vertical strips of bounded width. Thus we conclude from (4.6) and Lemma 4 that $U_\xi(z,s,\lambda)$ equals

$$\frac{2\sqrt{\pi}}{|t|^{1/2}} e\left(\lambda u_\xi - \frac{t}{8|t|}\right) \cdot \begin{cases} \operatorname{ch}\frac{v_\xi}{2}\left[\left(\operatorname{th}\frac{v_\xi}{2}\right)^{1-s} + \left(\operatorname{th}\frac{v_\xi}{2}\right)^{s} \frac{it}{|t|}(-1)^{2\pi\lambda}\right], & \text{if } \xi=\zeta , \\ \frac{1}{2}\, v_\xi^{1-s} + v_\xi^{s} \frac{it}{|t|}\left|\frac{\pi\lambda}{t}\right|^{2s-1} e^{2it} , & \text{if } \xi=\vartheta \text{ and } \lambda \neq 0 \\ \cos\frac{v_\xi}{2}\left[\left(\operatorname{tg}\frac{v_\xi}{2}\right)^{1-s} + \varepsilon\left(\operatorname{tg}\frac{v_\xi}{2}\right)^{s}\right] , & \text{if } \xi = (\eta_1,\eta_2,\varepsilon) , \end{cases} \tag{9.12}$$

up to an error $o(|t|^{-1/2})$, uniformly for s in $\frac{1}{2} \leqslant \sigma \leqslant 3/2$ provided that $v_\xi < \frac{\pi}{2}$ in case $\xi = \eta$.

Since $|V_\xi(z,s,\lambda)| \leqslant |V_\xi(z,\sigma,o)|$ in $\sigma > 0$ by (4.2) and (4.8), Proposition 1 and (9.8) yield

$$ {}_\xi P_\chi(s,m,n)U_\chi(z,s,\frac{n}{\lambda_\chi}) << \int_o^{\lambda_\chi} |S_\chi^{(1)} P_\xi^\psi(M_\chi^{-1}\exp(\varrho X_\chi)(z),s,m)|d\varrho + O(1) \qquad (9.13) $$

for z in $\mathfrak{U}_\chi$ and uniformly for s in $\frac{1}{2} \leqslant \sigma \leqslant 3/2$.

If $\mathfrak{R}_T$ denotes the rectangle with vertices at $2+i(T-1)$, $2+i(2T+1)$, $\frac{1}{4}+i(2T+1)$ and $\frac{1}{4}+i(T-1)$ in that order Lemma 13 and Cauchy's theorem show that

$$ \int_T^{2T}\int_{1/2}^{3/2} |\Psi_\xi(s,s',m)|d\sigma dt << \int_{\mathfrak{R}_T} |\psi_\xi(w,s',m) \int_T^{2T}\int_{1/2}^{3/2} \frac{d\sigma dt}{|(s-\bar{s}')(s-1+\bar{s}')(w-s)|} |dw| $$

$$ << \frac{T\log T}{|s'|^2+T^2}\int_{\mathfrak{R}_T} |\psi_\xi(w,s',m)||dw| \ , \quad T \to \infty \ , $$

uniformly for $s' = \sigma' + it'$ in $\frac{1}{2} \leqslant \sigma' \leqslant 1$. Thus it follows from Proposition 2(ii), Cauchy's inequality, (9.5) and (9.6) that

$$ \int_T^{2T}\int_{1/2}^{3/2} |P_\xi^\psi(z,s,m)|d\sigma dt << T\log T \int_{\mathfrak{R}_T} \Big\{ \sum_{j\geqslant o} \frac{|\alpha_{j\xi}(m)e_j(z)|}{|s_j|^2+T^2} |\psi_\xi(w,s_j,m)| + $$

$$ + \frac{1}{4\pi i}\sum_{\iota=1}^{\kappa}\int_{(1/2)} \Big|\alpha_{\iota\xi}(s'm)\psi_\xi(w,s',m) + \delta(\xi,\vartheta_\iota)\delta_{mo}\frac{\psi_\xi(w,1-s',m)}{B(1/2,s'-1/2)}\Big| \qquad (9.14) $$

$$ \frac{|E_\iota(z,s')|}{|s'|^2+T^2}ds'\Big\}|dw| << T\log T\int_{\mathfrak{R}_T} <\tilde{P}_\xi(\cdot,w,m),\tilde{P}_\xi(\cdot,w,m)>^{1/2}|dw| \cdot $$

$$ \cdot <P_z(\cdot,2,0),P_z(\cdot,2,0)>^{1/2} << T^3\log T $$

locally uniformly for z in $\mathfrak{H}$. Clearly the same bound holds if we replace s by $\bar{s}$ in the first integral of (9.14). Therefore we obtain the lemma from (9.12)-(9.14) by a simple measure theoretical argument.

10. Sum formulae (first form)

In this section we derive identities relating spectral data of the Laplacian in $L_2(\Gamma\backslash\mathfrak{H})$ to algebraic data of Γ. These identities enable us to study the Fourier coefficients $\alpha_{j\xi}(m)$ and $\alpha_{\iota\xi}(s,m)$ on average over the spectrum in terms of the generalized Kloosterman sums.

Theorem 1

(i) Let $g(s)$ be analytic for s in $-\frac{1}{2}\leq\sigma\leq 3/2$ such that

$$h(s) = g(s)\gamma_\xi(s,\lambda) + g(1-s)\gamma_\chi(1-s,\lambda') << |s|^{-B}$$

uniformly in $\frac{1}{2}\leq\sigma\leq 3/2$ with a $B>2$. In the notation of section 6 let

$${}_\xi h_\chi(\lambda,\lambda',\nu) =$$

$$\sum_{\delta,\delta'=0}^{1}\varepsilon^\delta\varepsilon'^{\delta'}\int_{{}_\xi^\delta\mathcal{C}_\chi^{\delta'}} e(-\lambda\tau+\lambda'\operatorname{Re}{}_\xi\varphi_\chi(\tau+\delta\pi i,\nu))\frac{1}{4\pi^2 i\beta_\xi}\cdot$$

$$\cdot\int_{(1/2)} h(s)(s-1/2)\operatorname{ctg}\pi s|{}_\xi q_\chi(\tau+\delta\pi i,\nu)|^{-s}ds\,d\tau$$

and let ${}_\xi H_\chi(\lambda,\lambda',\nu)$ be defined by the right hand side of the preceding equation with ${}_\xi^\delta\mathcal{C}_\chi^{\delta'}$, ${}_\xi\Phi_\chi$ and ${}_\xi Q_\chi$ in place of ${}_\xi^\delta\mathcal{C}_\chi^{\delta'}$, ${}_\xi\varphi_\chi$ and ${}_\xi q_\chi$ respectively. If $\lambda=\frac{m}{\lambda_\xi}$, $\lambda'=\frac{n}{\lambda_\chi}$ and if $h(s)\gamma_\xi(1-s,\lambda)$ is analytic at $s=1$ then

$$\frac{1}{4\pi i}\int_{(1/2)}\sum_{\iota=1}^{\kappa}\bar{\alpha}_{\iota\xi}(s,m)\alpha_{\iota\chi}(s,n)h(s)ds +$$

$$+\sum_{t_j\geq 0}\bar{\alpha}_{j\xi}(m)\alpha_{j\chi}(n)h(s_j) + \sum_{\frac{1}{2}<s_j\leq 1}\bar{\alpha}_{j\xi}(m)\alpha_{j\chi}(n)h(s_j)\gamma_\xi(1-s_j,\lambda) =$$

$$=\frac{\delta(\xi,\chi)\delta_{mn}+\delta^*(\xi,\chi)\delta_{-mn}\varepsilon e(\frac{m}{\lambda_\xi}\varrho_\xi)}{4\pi^2 i\beta_\xi\lambda_\xi}\int_{(1/2)}(s-1/2)\operatorname{ctg}\pi s\, h(s)ds$$

$$+\frac{1}{\lambda_\xi\lambda_\chi}\sum_\nu{}_\xi s_\chi(m,n,\nu)\,{}_\xi h_\chi(\lambda,\lambda',\nu) +$$

$$+\frac{1}{\lambda_\xi\lambda_\chi}\sum_{\delta,\delta'=0}^{1}\varepsilon^\delta\varepsilon'^{\delta'}\sum_\nu{}_\xi^\delta S_\chi^{\delta'}(m,n,\nu)\;{}_\xi H_\chi((-1)^\delta\lambda,(-1)^{\delta'}\lambda',\nu)\quad ,$$

unless $\xi = \vartheta$, $m = 0$ or $\chi = \vartheta'$, $n = 0$.

(ii) Let $g(s)$ and $h(s)$ be analytic in $-\frac{1}{2} \leq \sigma \leq 3/2$ such that

$$g(s)\ h(1/2 \pm (s-1/2)) << |s|^{-B}$$

uniformly in $\frac{1}{2} \leq \sigma \leq 3/2$ with a $B > 2$. If $h(s)\,\gamma_\chi(s,\frac{n}{\lambda_\chi})$ is analytic at $s = 0$ then

$$\frac{1}{4\pi i}\int_{(1/2)} \sum_{\iota=1}^{\kappa} \Big(\delta(\vartheta,\vartheta_\iota)\,\frac{g(s)}{B(1/2,s-1/2)} +$$

$$\bar{\alpha}_{\iota\vartheta}(s,0)\,g(1-s)\Big)\Big(\delta(\chi,\vartheta_\iota)\,\delta_{0n}\,\frac{h(1-s)}{B(1/2,1/2-s)} + \alpha_{\iota\chi}(s,n)\,h(s)\Big)\,ds$$

$$+ \frac{1}{2}\Big(1+\delta_{0n}\sum_{\iota=1}^{\kappa}\delta(\chi,\vartheta_\iota)\Big) \sum_{\frac{1}{2}<s_j\leq 1} \bar{\alpha}_{j\vartheta}(0)\,\alpha_{j\chi}(n)\,g(s_j)\,\{h(s_j) + h(1-s_j)\,\gamma_\chi(1-s_j,\frac{n}{\lambda_\chi})\}$$

$$= \frac{\delta(\vartheta,\chi)\,\delta_{0n}}{2\pi^2 i}\int_{(1/2)} (s-1/2)\,\mathrm{ctg}\,\pi s\ g(s)\,h(1-s)\,ds +$$

$$\sum_{\delta'=0}^{1} \varepsilon'^{\delta'} \sum_{\nu} {}^{0}_{\vartheta}S_{\chi}^{\delta'}(0,n,\nu)\ \frac{1+\delta_{0n}\sum_{\iota=1}^{\kappa}\delta(\chi,\vartheta_\iota)}{4\pi i\lambda_\chi}\ \cdot$$

$$\cdot \int_{(1/2)} \frac{\nu^{-2s} g(s)}{B(1/2,s-1/2)}\,\{h(s) + h(1-s)\,\gamma_\chi(1-s,\frac{n}{\lambda_\chi})\}\,ds \quad .$$

<u>Proof</u>: First assume that ξ , $m = \lambda\lambda_\xi$ and χ , $n = \lambda'\lambda_\chi$ are among the cases considered in part (i). It follows from Proposition 3(ii) that

$$\frac{1}{2\pi i}\Big(\int_{\mathcal{L}_+} - \int_{\mathcal{L}_-}\Big)(2s-1)\,{}_{\xi}P_{\chi}(s,m,n)\,g(s)\,ds = 2\pi\beta_\xi\lambda_\xi \sum_{t_j\geq 0} \bar{\alpha}_{j\xi}(m)\,\alpha_{j\chi}(n)\,h(s_j)\ , \quad (10.1)$$

where $\mathcal{L}_\pm$ are as in the proof of Proposition 2 (ii). The definition of these paths, Lemma 4 and Proposition 3(iii) show that after simple shifts and substitutions the left hand side of (10.1) equals

$$\frac{1}{2\pi i}\int_{\mathscr{C}_+}(2s-1)\{{}_\xi P_\chi(s,m,n)g(s) + {}_\xi P_\chi(1-s,m,n)g(1-s)\}ds =$$

$$\frac{1}{2\pi i}\int_{\mathscr{C}_+}(2s-1)\,{}_\xi P_\chi(s,m,n)h(s)\gamma_\xi(1-s,\lambda)\,ds + \tag{10.2}$$

$$+\frac{1}{2\pi i}\int_{(1/2)}(2s-1)\operatorname{ctg}\pi s\,\frac{g(1-s)}{\gamma_\chi(s,\lambda')}\,[\delta(\xi,\chi)\delta_{mn} + \delta^*(\xi,\chi)\delta_{-mn}\varepsilon e(\tfrac{m}{\lambda_\xi}\varrho_\xi)$$

$$+ {}_\xi p_\chi(s,m,n)\gamma_\xi(1-s,\lambda)]ds - \frac{\beta_\xi\lambda_\xi}{i}\int_{(1/2)}\sum_{\iota=1}^{\kappa}\alpha_{\iota\xi}(1-s,-m)\alpha_{\iota\chi}(s,n)\frac{g(1-s)}{\gamma_\chi(s,\lambda')}ds\,.$$

We infer from Lemma 4, (7.10), Proposition 3(i) and the first equation of Proposition 3(iv) that we may replace $g(1-s)/\gamma_\chi(s,\lambda')$ by $\frac{1}{2}h(s)$ in the last two integrals. Moreover, Lemma 15 permits to shift the contour in the first term on the right of (10.2). Since $h(s)\gamma_\xi(1-s,\lambda)$ is analytic in $\frac{1}{2}\leqslant\sigma\leqslant 3/2$ we thus obtain from (7.19), Proposition 3 (ii), (10.1) and (10.2)

$$\frac{1}{4\pi i}\int_{(1/2)}\sum_{\iota=1}^{\kappa}\bar{\alpha}_{\iota\xi}(s,m)\alpha_{\iota\chi}(s,n)h(s)ds + \sum_{t_j\geqslant 0}\bar{\alpha}_{j\xi}(m)\alpha_{j\chi}(n)h(s_j) +$$

$$+\sum_{\frac{1}{2}<s_j\leqslant 1}\bar{\alpha}_{j\xi}(m)\alpha_{j\chi}(n)h(s_j)\gamma_\xi(1-s_j,\lambda) \tag{10.3}$$

$$= \frac{1}{4\pi^2 i\beta_\xi\lambda_\xi}\int_{(1/2)}(s-1/2)\operatorname{ctg}\pi s\,h(s)[\delta(\xi,\chi)\delta_{mn} + \delta^*(\xi,\chi)\delta_{-mn}\varepsilon e(\tfrac{m}{\lambda_\xi}\varrho_\xi) +$$

$$+ {}_\xi p_\chi(s,m,n)\gamma_\xi(1-s,\lambda)]ds + \frac{1}{4\pi^2 i\beta_\xi\lambda_\xi}\int_{(3/2)}(2s-1)\,{}_\xi P_\chi(s,m,n)h(s)\gamma_\xi(1-s,\lambda)\,ds,$$

provided that $h(s)$ decays more rapidly than any inverse power of $|s|$ in $\frac{1}{2}\leqslant\sigma\leqslant 3/2$. Now ${}_\xi p_\chi(s,m,n)$ is bounded on $\sigma = 1/2$ by Lemma 6 (ii), Lemma 11 and (7.11), while ${}_\xi P_\chi(s,m,n)$ is bounded on $\sigma = 3/2$ by Lemma 9 since the series in (7.12) converges absolutely in $\sigma > 1$. Therefore all terms on the right of (10.3) still make sense if only h decays as assumed in Theorem 1(i). Thus we conclude via a simple limiting argument that (10.3) continues to hold for all such h .

If we insert (7.11) and (7.12) for ${}_{\xi}p_{\chi}$ and ${}_{\xi}P_{\chi}$ their corresponding contributions to the right of (10.3) amount to

$$\frac{1}{4\pi^2 i\beta_{\xi}\lambda_{\xi}\lambda_{\chi}} \sum_{\nu} {}_{\xi}s_{\chi}(m,n,\nu) \int_{(1/2)} (s-1/2)\operatorname{ctg}\pi s\ h(s)\gamma_{\xi}(1-s,\lambda)\ {}_{\xi}i_{\chi}(\nu,s,\lambda,\lambda')ds \qquad (10.4)$$

and

$$\frac{1}{4\pi^2 i\beta_{\xi}\lambda_{\xi}\lambda_{\chi}} \sum_{\delta,\delta'=0}^{1} \varepsilon^{\delta}\varepsilon'^{\delta'} \sum_{\nu} {}_{\xi}^{\delta}S_{\chi}^{\delta'}(m,n,\nu) \cdot$$
$$\int_{(3/2)} (2s-1)h(s)\gamma_{\xi}(1-s,\lambda)\ {}_{\xi}J_{\chi}(\nu,s,(-1)^{\delta}\lambda,(-1)^{\delta'}\lambda')ds \qquad (10.5)$$

respectively. In (10.5) we can shift the integration back to the $\frac{1}{2}$ - line without picking up anything from residues. Now note that

$$h(s)\gamma_{\xi}(1-s,\lambda) = g(s) + g(1-s)\gamma_{\xi}(1-s,\lambda)\gamma_{\chi}(1-s,\lambda') = h(1-s)\gamma_{\chi}(1-s,\lambda')$$

by Lemma 4, whence

$$\int_{(1/2)} (2s-1)h(s)\gamma_{\xi}(1-s,\lambda)\ {}_{\xi}J_{\chi}(\nu,s,\lambda,\lambda')ds =$$
$$\int_{(1/2)} (s-1/2)[h(s)\gamma_{\xi}(1-s,\lambda)\ {}_{\xi}J_{\chi}(\nu,s,\lambda,\lambda') -$$
$$h(1-s)\gamma_{\xi}(s,\lambda)\ {}_{\xi}J_{\chi}(\nu,1-s,\lambda,\lambda')]ds =$$
$$\int_{(1/2)} (s-1/2)\operatorname{ctg}\pi s\ h(s)\gamma_{\chi}(s,\lambda')\ {}_{\xi}I_{\chi}(\nu,1-s,\lambda,\lambda')ds$$

by the first functional equation of Lemma 10. Hence the integral representations given in Lemma 10 and 11 imply that the integrals in (10.4) and (10.5) are equal to

$$4\pi^2 i\beta_{\xi\xi}h_{\chi}(\lambda,\lambda',\nu) \quad \text{and} \quad 4\pi^2 i\beta_{\xi\xi}H_{\chi}(-1)^{\delta}\lambda,(-1)^{\delta'}\lambda',\nu)$$

respectively. For the so arising double integrals are absolutely convergent unless $\xi = \vartheta$ or $\chi = \vartheta'$. In the remaining cases we easily justify the interchange of integrations by using the absolutely convergent representations of ${}_{\xi}I_{\chi}$ given in the proof of Lemma 10. Thus part (i) follows from (10.3) - (10.5).

By (7.18), (7.19) and (9.10) the first term of the identity in part (ii) can be written as

$$\frac{\delta(\vartheta,\chi)\delta_{on}}{4\pi^2 i}\int_{(1/2)}(s-1/2)\,\mathrm{ctg}\,\pi s\; g(s)h(1-s)\,ds\;+$$

$$\frac{1}{4\pi i}\int_{(1/2)}\frac{{}_{\vartheta}P_{\chi}(s,o,n)\,g(s)h(s)}{B(1/2,s)B(1/2,s-1/2)}\,ds\;+$$

$$\frac{\delta_{on}}{4\pi i}\sum_{\iota=1}^{\kappa}\delta(\chi,\vartheta_{\iota})\int_{(1/2)}\frac{{}_{\vartheta}P_{\chi}(1-s,o,o)}{B(1/2,1-s)B(1/2,1/2-s)}\,g(1-s)h(1-s)\,ds\;+$$

$$\frac{1}{4\pi i}\int_{(1/2)}\sum_{\iota=1}^{\kappa}\bar{\alpha}_{\iota\vartheta}(s,o)\,\alpha_{\iota\chi}(s,n)\,g(1-s)h(s)\,ds\;. \tag{10.6}$$

If we substitute s by $1-s$ in the last two terms and insert the second functional equation of Proposition 3(iv) in the last term (10.6) transforms into

$$\frac{\delta(\vartheta,\chi)\delta_{on}}{2\pi^2 i}\int_{(1/2)}(s-1/2)\,\mathrm{ctg}\,\pi s\; g(s)h(1-s)\,ds\;+$$

$$\frac{1+\delta_{on}\sum_{\iota=1}^{\kappa}\delta(\chi,\vartheta_{\iota})}{4\pi i}\int_{(1/2)}\frac{{}_{\vartheta}P_{\chi}(s,o,n)}{B(1/2,s)B(1/2,s-1/2)}\,g(s)\,\{h(s)+h(1-s)\gamma_{\chi}(1-s,\tfrac{n}{\lambda_{\chi}})\}\,ds\;. \tag{10.7}$$

Since $h(s)\gamma_{\chi}(s,\frac{n}{\lambda_{\chi}})$ is analytic in $-\frac{1}{2}\leq\sigma\leq 1/2$ we deduce from (4.19), (5.2) and Proposition 3(ii) that the last integral is equal to

$$-2\pi i\sum_{\frac{1}{2}<s_j\leq 1}\bar{\alpha}_{j\vartheta}(o)\,\alpha_{j\chi}(n)\,g(s_j)\{h(s_j)+h(1-s_j)\gamma_{\chi}(1-s_j,\tfrac{n}{\lambda_{\chi}})\}$$

$$+\int_{(3/2)}\frac{{}_{\vartheta}P_{\chi}(s,o,n)\,g(s)}{B(1/2,s)B(1/2,s-1/2)}\{h(s)+\gamma_{\chi}(1-s,\tfrac{n}{\lambda_{\chi}})h(1-s)\}\,ds\;.$$

Again it is permissible to plug the series (7.12) into this integral over the $\frac{3}{2}$ - line and to interchange summation with integration. Thus by Lemma 8 and a shift back to the 1/2 - line the preceding integral takes the form

$$\frac{1}{\lambda_\chi} \sum_{\delta'=0}^{1} \varepsilon'^{\delta'} \sum_{\nu} {}^{0}S_{\vartheta\chi}^{\delta'}(0,n,\nu) \int_{(1/2)} \frac{\nu^{-2s} g(s)}{B(1/2,s-1/2)} \{h(s) + h(1-s)\gamma_\chi(1-s, \tfrac{n}{\lambda_\chi})\} ds$$

Hence the identity of part (ii) is established.

Strictly speaking, g or h have to decay rapidly enough to justify the shift from the 1/2- to the 3/2-line by Lemma 15. On the other hand, part(i) and the functional equation in the remark following Proposition 3 show that all integrals we have written down converge absolutely under the assumptions of part (ii). Thus a limiting argument again allows to uphold the identity under those assumptions and the theorem follows.

Remark. Theorem 1 (i) can also be given a form in which all s_j enter on equal terms. For by Lemma 4 there are functions $\Gamma_\xi(s,\lambda)$ such that

$$\gamma_\xi(s,\lambda) = \frac{\Gamma_\xi(s,\lambda)}{\Gamma_\xi(1-s,\lambda)} \quad \text{and} \quad \overline{\Gamma}_\xi(s,\lambda) = \Gamma_\xi(\bar{s},\lambda) \ ,$$

unless $\xi = \vartheta$, $\lambda = 0$.

If

$$\tilde{h}(s) = h(s)\,\Gamma_\xi(1-s,\lambda)\,\Gamma_\chi(s,\lambda')$$

we then note that

$$\tilde{h}(1-s) = h(s)\frac{\gamma_\xi(1-s,\lambda)}{\gamma_\chi(1-s,\lambda')}\,\Gamma_\xi(s,\lambda)\,\Gamma_\chi(1-s,\lambda') = \tilde{h}(s)$$

by the functional equation of h and that

$$\frac{\tilde{h}(s)}{\overline{\Gamma}_\xi(s,\lambda)\,\Gamma_\chi(s,\lambda)} = \begin{cases} h(s) & \text{on} \quad \sigma = 1/2 \ , \\ h(s)\,\gamma_\xi(1-s,\lambda) & \text{on} \quad t = 0 \ . \end{cases}$$

In our search for a unified approach to Theorem 1 we were led to work with integral representations. So far there was no need to identify them with the special functions they represent. This complicates a direct comparison with existing results in the few cases previously considered by other authors. The reader may also wish to see what Theorem 1 yields in the most familiar situations. For these reasons we reformulate two cases of Theorem 1 in more traditional terms. The first of them arises with $\xi = \vartheta$ and $\chi = \vartheta$. It has found many applications to analytic number theory in recent years. In particular our results of [7], [8], [9] rely on it. If moreover $mn \neq 0$, it contains results of Bruggeman [1] and Kuznietsov[13]. The second case deals with the spectral kernel function of the Laplacian in $L_2(\Gamma\backslash\mathfrak{H})$ which has attracted the interest of many people for a long time. We are not aware, however, that Corollary 2 has previously been stated in our form.

Corollary 1. Let Γ contain matrices of the form $\pm\exp(\tau X_{\vartheta})$ precisely if τ is an integer. If $\vartheta_1 = \infty$ and $K_\nu(z)$ is the modified Bessel function in its usual notation (cf. [14] p. 66) then

$$e_j(z) = \alpha_{jo}y^{1-s_j} + \sum_{n\neq o}\alpha_{jn}y^{1/2}K_{s_j-1/2}(2\pi|n|y)e(nx) \quad , \; j \geq 0 \; ,$$

and

$$E_\iota(z,s) = \delta_{\iota 1}y^s + \alpha_{\iota o}(s)y^{1-s} +$$

$$+ \sum_{n\neq o}\alpha_{\iota n}(s)y^{1/2}K_{s-1/2}(2\pi|n|y)e(nx)\, , \quad \iota = 1,\dots,\kappa \; ,$$

where the coefficients α_{jn} and $\alpha_{\iota n}(s)$ are related to $S(m,n,\nu) = {}^{o}_{\infty}S^{o}_{\infty}(m,n,\nu)$ by the following identities: For $mn \neq 0$

$$\frac{1}{4\pi i}\int_{(1/2)}\sum_{\iota=1}^{\kappa}\bar{\alpha}_{\iota m}(s)\alpha_{\iota n}(s)g(s)\,ds + \sum_{j\geq o}\bar{\alpha}_{jm}\alpha_{jn}g(s_j)$$

$$= \frac{\delta_{mn}}{\pi^2 i}\int_{(1/2)}(s-1/2)\cos\pi s\; g(s)\,ds + \sum_{\nu} S(m,n,\nu)H(m,n,\nu)$$

with

$$H(m,n,\nu) = \frac{2}{\pi^2 i\nu}\int_{o}^{\infty}\cos\left(\frac{2\pi}{\nu}|mn|^{1/2}\left[\varrho + \frac{mn}{|mn|\varrho}\right]\right)\cdot$$

$$\int_{(1/2)}(s-1/2)\cos\pi s\; g(s)\varrho^{-2s}\,ds\,d\varrho \; ,$$

while

$$\frac{1}{4\pi i}\int_{(1/2)}\sum_{\iota=1}^{\kappa}\left(\delta_{\iota 1}g(s)+\bar{\alpha}_{\iota o}(s)g(1-s)\right)\left(\delta_{\iota 1}\delta_{on}h(1-s)+\alpha_{\iota n}(s)h(s)\right)ds$$

$$+\frac{1+\delta_{on}}{2}\sum_{\frac{1}{2}<s_j\leq 1}\bar{\alpha}_{jo}\alpha_{jn}g(s_j)\{h(s_j)+h(1-s_j)\}=\frac{\delta_{on}}{2\pi i}\int_{(1/2)}g(s)h(1-s)ds$$

$$+\sum_{\nu}S(o,n,\nu)\frac{1+\delta_{on}}{4\pi i}\int_{(1/2)}\nu^{-2s}g(s)\{h(s)+h(1-s)\}A_n(s)ds ,$$

with

$$A_n(s)=\begin{cases}\dfrac{2\pi^s|n|^{s-1/2}}{\Gamma(s)} & , \text{ if } n\neq 0 ,\\ B(1/2,s-1/2) & , \text{ if } n=0 .\end{cases}$$

In these identities $g(s)$ and $h(s)$ have to be analytic in $-\frac{1}{2}\leqslant\sigma\leqslant 3/2$ such that in the first

$$g(s) = g(1-s) << e^{-\pi|s|}|s|^{-B}$$

and in the second

$$g(s)h(\tfrac{1}{2}\pm(s-1/2)) << \frac{|s|^{1/2-B}}{|A_n(1/2\pm(s-1/2))|}$$

uniformly in $\frac{1}{2}\leqslant\sigma\leqslant 3/2$ with a $B>2$.

<u>Proof</u>: By our assumptions , Γ has a cusp at infinity and the identity matrix satisfies the conditions (5.1) and (5.2) for M_∞ . We deduce from (3.6) , (4.6) , (4.7) and the last formula on p. 85[14] that

$$U_\vartheta(z,s,\lambda) = 2e(\lambda x)y^s\int_0^\infty (y^2+\varrho^2)^{-s}\cos(2\pi|\lambda|\varrho)\,d\varrho =$$

$$= y^{1/2}K_{s-1/2}(2\pi|\lambda|x)\,e(\lambda x)\ \frac{2|\lambda|^{s-1/2}}{\pi^{-s}\Gamma(s)}$$

if $\lambda\neq 0$. Thus (4.15) , Lemma 5 and (7.17) yield the following relations between the old and new Fourier coefficients of e_j and E_ι :

$$\alpha_{jn} = \alpha_{j\infty}(n)A_n(s_j) ,\ j\geqslant 0 ,\ \alpha_{\iota n}(s) = \alpha_{\iota\infty}(s,n)A_n(s) ,\ \iota = 1,\ldots,\kappa . \tag{10.8}$$

Hence we obtain from Lemma 4

$$\sum_{j\geq 0} \bar{\alpha}_{jm}\alpha_{jn} g(s_j) = \sum_{t_j\geq 0} \bar{\alpha}_{j\infty}(m)\,\alpha_{j\infty}(n)\,h(s_j) +$$

$$+ \sum_{\frac{1}{2}<s_j\leq 1} \bar{\alpha}_{j\infty}(m)\,\alpha_{j\infty}(n)\,h(s_j)\,\gamma_\infty(1-s_j,m)$$

if $mn \neq 0$ and

$$h(s) = 4g(s)\left|\frac{n}{m}\right|^{s-1/2}\sin\pi s \; .$$

Thus the left hand sides of the first identity in Theorem 1 and in Corollary 1 agree if we use Theorem 1(i) with $\xi = \chi = \infty$ and h as above. This h also satisfies the assumptions of Theorem 1(i) in this case since

$$2h(s) = h(s)\,\gamma_\infty(1-s,m)\,\gamma_\infty(s,m) + h(1-s)\,\gamma_\infty(s,m)\,\gamma_\infty(1-s,n)$$

by Lemma 4. Moreover the first term on the right of those two identities agree by (4.19), (5.2) and (7.10). The second term on the right of Theorem 1(i) vanishes by (5.8) and (5.11) since ${}_\infty g_\infty = \emptyset$ by (3.9), (3.10) and (3.12). Finally, only the summand with $\delta = \delta' = 0$ gives a contribution to the last term in Theorem 1(i) in view of Lemma 1(i), (5.7) and (5.10) while (6.14), (6.16), (6.17) and Lemma 10 yield

$${}_\infty H_\infty(m,n,\nu) = \int_{-\infty}^{\infty} e\left(-m\tau - \frac{n}{\nu^2\tau}\right)\frac{1}{\pi^2 i}\int_{\left(\frac{1}{2}\right)} g(s)\left|\frac{n}{m}\right|^{s-1/2}(s-1/2)\cos\pi s\,|\nu\tau|^{-2s}\,ds\,d\tau$$

$$= H(m,n,\nu)$$

by the substitution $\tau = \varrho\left|\frac{n}{m}\right|^{1/2}/\nu$. This establishes the first identity of Corollary 1.

The second identity immediately follows if in Theorem 1(ii) we choose $\xi = \chi = \infty$, substitute (10.8) and take $g(s)A_o(s)$, $h(s)A_n(s)$ in place of $g(s)$, $h(s)$ respectively.

<u>Remark</u>: Corollary 1 contains all identites we required in [7], [8], [9] and promised to prove later. For the top formula on p.81 [14] and the equation $g(s) = g(1-s)$ show that

$$H(m,n,\nu) = \frac{4}{\pi^2 i\nu} \int_{(1/2)} g(s)(s-1/2)\cos\pi s \int_0^\infty \cos\left(\frac{4\pi}{\nu}|mn|^{1/2}\operatorname{ch}\varrho\right)\operatorname{ch}((2s-1)\varrho)\,d\varrho\,ds$$

$$= \frac{1}{\pi i\nu} \int_{(1/2)} g(s)(2s-1)J_{2s-1}\left(\frac{4\pi}{\nu}|mn|^{1/2}\right)ds \quad , \quad \text{if} \quad mn>0 \ , \tag{10.9}$$

where $J_\nu(z)$ is the Bessel function (cf. [14] p. 65). In [7] Lemma 2 we erroneously stated (10.9) for $mn \neq 0$. However this does not affect the results of [7].

<u>Corollary 2</u>. Let $h(s)$ be analytic in $-\frac{1}{2} \leqslant \sigma \leqslant 3/2$ such that

$$h(s) = h(1-s) \ll |s|^{-B}$$

uniformly in $\frac{1}{2} \leqslant \sigma \leqslant 3/2$ with a $B<2$. If ζ,ζ' are arbitrary points in $\mathfrak{H}$

$$\frac{1}{4\pi i} \int_{(1/2)} \sum_{i=1}^{\kappa} \bar{E}_\iota(\zeta,s)E_\iota(\zeta',s)h(s)\,ds + \sum_{j\geqslant 0} \bar{e}_j(\zeta)e_j(\zeta')h(s_j)$$

$$= \delta(\zeta,\zeta')[\Gamma_\zeta:Z_\Gamma]H(1) + [\Gamma_\zeta:Z_\Gamma][\Gamma_{\zeta'}:Z_\Gamma] \sum_\nu {}_\zeta^{o}S_{\zeta'}^{o}(o,o,\nu)H(\nu) \ ,$$

where

$$H(\nu) = \frac{1}{8\pi^2 i} \int_0^{2\pi} \int_{(1/2)} h(s)(s-1/2)\operatorname{ctg}\pi s\left(\left(\nu \sin\frac{\tau}{2}\right)^2 + \left(\frac{\cos\frac{\tau}{2}}{\nu}\right)^2\right)^{-s} ds\,d\tau .$$

<u>Proof</u>: By (4.21), (4.24), (5.1), Lemma 5 and (7.17) we have

$$e_j(\zeta) = 2\pi\alpha_{j\zeta}(o) \ , \ j \geqslant 0 \ , \text{ and } \ E_\iota(\zeta,s) = 2\pi\alpha_{\iota\zeta}(s,o) \ ,$$
$$\iota = 1,\dots,\kappa \ .$$

We now apply Theorem 1(i) with $\xi = \zeta$, $m = o$ and $\chi = \zeta'$, $n = o$. In this situation the h of Corollary 2 qualify for Theorem 1(i) since $\gamma_\zeta(s,o) = 1$ by Lemma 4. As in Corollary 1 we observe that in Theorem 1(i) the second term on the right vanishes while there is a contribution to the last term only if $\delta = \delta' = o$. Moreover (4.19), (6.6), (6.14), (6.17) and Lemma 10 yield ${}_\zeta H_{\zeta'}(o,o,\nu) = H(\nu)$. Thus

Corollary 2 follows from (5.2) and (7.10).

Theorem 2. For all ξ and m we have

$$\frac{1}{4\pi}\int_{-T}^{T}\sum_{\iota=1}^{\kappa}|\alpha_{\iota\xi}(\tfrac{1}{2}+it,m)|^2dt+\sum_{0\le t_j\le T}|\alpha_{j\xi}(m)|^2=\frac{\delta(\xi,\xi)+\varepsilon\delta^*(\xi,\xi)\delta_{mo}}{4\pi^2\beta_\xi\lambda_\xi}T^2$$

$+ O(T)$, $T \to \infty$,

in the notation of (4.19), (5.2) and (7.10) .

Proof: By the remark following Proposition 3, Theorem 2 obviously holds for $\xi = \vartheta$, $m = 0$ so that we can apply Theorem 1(i) in the remaining cases with $\xi = \chi$, $m = n$. If we choose

$$g(s) = \frac{s(1-s)}{i\gamma_\xi(s,\lambda)}\left(T\int_{\frac{s-1/2}{T}}^{i\infty}e^{\pi T^2(z-i)^2}dz-1/2\right)$$

with a parameter $T \geq 1$ then

$$h(s) = g(s)\gamma_\xi(s,\lambda) + g(1-s)\gamma_\xi(1-s,\lambda) = \frac{s(1-s)}{i}\int_{s-1/2-iT}^{s-1/2+iT}e^{\pi\tau^2}d\tau ,$$

whence $h(s)$ satisfies all assumptions of Theorem 1(i). Since

$$\int_t^\infty e^{-\tau^2}d\tau \leq e^{-t^2}$$

for $t > 0$ we have

$$h(1/2+it) = (t^2+1/4)\left\{\chi_{(-1,1)}(\tfrac{t}{T}) + O\left(e^{-\pi(t-T)^2}+e^{-\pi(t+T)^2}\right)\right\} \tag{10.10}$$

uniformly in T , where $\chi_{(-1,1)}$ denotes the characteristic function of the interval $(-1,1)$.

Now suppose that for the $h(s)$ above

$$_\xi h_\xi(\lambda,\lambda,\nu) \ll T^3 \quad\text{and}\quad {}_\xi H_\xi(\lambda,\lambda,\nu) \ll T^3\nu^{-\sigma_2} \tag{10.11}$$

with a $\sigma_2 > 2$ and that the same bounds also hold for the $h(s)$ corresponding to

$$g(s) = \frac{s(1-s)}{\gamma_\xi(s,\lambda)}\left(e^{\pi(s-1/2-iT)^2} + e^{\pi(s-1/2+iT)^2}\right) .$$

Then we conclude from Lemma 9, Theorem 1(i) and (10.10) that

$$\frac{1}{4\pi}\int_{-T}^{T}\sum_{\iota=1}^{\kappa}|\alpha_{\iota\xi}(1/2+it,m)|^2\left(t^2+1/4\right)dt + \sum_{0\leq t_j\leq T}|\alpha_{j\xi}(m)|^2(t_j^2+1/4) =$$

$$\frac{\delta(\xi,\xi)+\varepsilon\delta^*(\xi,\xi)\,\delta_{mo}}{4\pi^2\beta_\xi\lambda_\xi}\int_{-T}^{T} t(t^2+1/4)\,\mathrm{th}\,\pi t\,dt + O\left(\int_0^\infty t^3 e^{-\pi(t-T)^2}dt + T^3\right),$$

since the sum in (7.11) is finite and the sum in (7.12) converges absolutely in $\sigma > 1$. Thus the above assumptions and integrations by parts imply Theorem 2. We shall only prove (10.11) since in every estimate below one verifies very similarly that the corresponding bound for our second choice of $g(s)$ is even by a factor T smaller than required.

If $f(s)$ is analytic on $\sigma = 0$ and sufficiently small near $\pm i\infty$ let $\hat{f}$ be given by

$$\hat{f}(\varrho) = \frac{1}{i}\int_{(0)} f(s)e^{-\varrho s}ds = \frac{1}{\varrho i}\int_{(0)} f'(s)e^{-\varrho s}ds .$$

In case

$$h_o(s) = \mathrm{tg}\,\pi s\,\frac{s(s^2-1/4)}{i}\int_{s-iT}^{s+iT} e^{\pi\tau^2}d\tau ,$$

we thus obtain from (10.10) by shifting the integration to the line $\sigma = \frac{\varrho}{|\varrho|}\sigma_1$

$$\hat{h}_o(\varrho) \ll e^{-|\varrho|\sigma_1}\int_0^\infty \{\min(t,\frac{1}{|\varrho|})t^2\chi_{(-1,1)}(\tfrac{t}{T}) + \min(1,\frac{1}{|\varrho|})t^3e^{-\pi(t-T)^2}\}dt$$

$$\ll e^{-|\varrho|\sigma_1}T^3\min(T,\frac{1}{|\varrho|})$$

for $0 < \sigma_1 < 3/2$. For $\nu > 1$ it follows from (6.14), (6.17) and the definition of the paths in Lemma 10 that

$$\left|\log|{}_\eta Q_\eta(\tau+\delta\pi i,\nu)|\right| \geq 2\log\nu$$

for τ on ${}^{\delta}_{\eta}\mathscr{C}^{\delta'}_{\eta}$ unless $\delta = \delta' = o$ and that

$${}_\xi Q_\xi(\tau,\nu) \ll (\nu\tau)^2 + \nu^{-2} \quad , \text{ if } \quad |\tau| \leq 1/\nu \ ,$$

$${}_\xi Q_\xi(\tau,\nu) \gg (\nu\tau)^2 \quad , \text{ if } \quad \min_{n\in Z}|\tau+n\operatorname{Re}\ell_\xi| \geq 1/\nu \ ,$$

for τ on ${}^{o}_{\xi}\mathscr{C}^{o}_{\xi}$. Let $\mathscr{C}_\xi(\nu)$ denote that part of the latter path on which $|\log{}_\xi Q_\xi(\tau,\nu)| \leq \frac{4}{5}\log\nu$. Since for our first choice of $g(s)$ the inner integrals in ${}_\xi H_\xi(\lambda,\lambda,\nu)$ are given by

$$i\,\hat{h}_o(\log|{}_\xi Q_\xi(\tau+\delta\pi i,\nu)|)\,|{}_\xi Q_\xi(\tau+\delta\pi i,\nu)|^{-1/2}$$

we conclude that

$${}_\xi H_\xi(\lambda,\lambda,\nu) - \frac{1}{4\pi^2\beta_\xi}\int_{\mathscr{C}_\xi(\nu)} e(-\lambda\tau+\lambda_\xi\Phi_\xi(\tau,\nu))\hat{h}_o(\log{}_\xi Q_\xi(\tau,\nu))\frac{d\tau}{{}_\xi Q_\xi^{1/2}(\tau,\nu)} \ll$$

$$\ll T^3\int_{\nu^{-3/5}}^{\infty}(\nu\tau)^{-1-2\sigma_1}d\tau + T^3\int_0^{\nu^{-7/5}}((\nu\tau)^2+\nu^{-2})^{\sigma_1-1/2}d\tau + T^3\nu^{-1-2\sigma_1} \ll$$

$$\ll T^3\nu^{-1-4\sigma_1/5} \tag{10.12}$$

On $\mathscr{C}_\xi(\nu)$ all integrands above behave alike. We readily infer from (6.14), (6.17) and (6.39) that

$$\frac{{}_\xi Q_\xi^{1/2}}{\partial_\xi Q_\xi/\partial\tau}(\tau,\nu) = \frac{\beta_\xi c_\xi(-\tau)}{2\nu|c_\xi(\tau)|} + O(\nu^{-2}[(\nu\tau)^2+(\nu\tau)^{-2}]) = \frac{\beta_\xi|c_\xi(-\tau)|}{2\nu|c_\xi(\tau)|} + O(\nu^{-6/5})$$
$$+ O(\nu^{-6/5})$$

and

$$\frac{\partial}{\partial\tau}\left(\frac{{}_\xi Q_\xi^{1/2}}{\partial_\xi Q_\xi/\partial\tau}(\tau,\nu)\right) \ll \nu^{-2}(|\nu\tau|+|\nu\tau|^{-3}) \ll \frac{\nu^{-6/5}}{|\nu\tau|}$$

for τ on $\mathscr{C}_\xi(\nu)$. If we set

$$h_1(s) = \frac{h_o(s-1/2)}{s-1/2} - \frac{h_o(s+1/2)}{s+1/2} \tag{10.13}$$

then by Cauchy's theorem

$$\hat{h}_1(\varrho) = \int_{(-1/2)} \frac{h_o(s)}{s} e^{-\varrho(s+1/2)} ds - \int_{(1/2)} \frac{h_o(s)}{s} e^{-\varrho(s-1/2)} ds =$$

$$= (e^{\varrho/2} - e^{-\varrho/2}) \int_{(o)} \frac{h_o(s)}{-s} e^{-\varrho s} ds ,$$

whence

$$\frac{d}{d\varrho}\left(\frac{\hat{h}_1(\varrho)}{e^{\varrho/2}-e^{-\varrho/2}}\right) = \hat{h}_o(\varrho) .$$

As for $\hat{h}_o$ one shows that

$$\hat{h}_1(\varrho)(e^{\varrho/2}-e^{-\varrho/2})^{-1} \ll e^{-|\varrho|\sigma_1} T^3$$

for $0 < \sigma_1 < 3/2$. Thus we obtain from (6.39) and (10.12) by partial integration

$$_\xi H_\xi(\lambda,\lambda,\nu) = \frac{1}{4\pi^2\beta_\xi} \int_{\mathcal{C}_\xi(\nu)} \frac{\hat{h}_1(\log{}_\xi Q_\xi(\tau,\nu))}{({}_\xi Q_\xi^{1/2}(\tau,\nu) - {}_\xi Q_\xi^{-1/2}(\tau,\nu))}$$

$$\cdot \frac{\partial}{\partial\tau}\left\{ e(-\lambda\tau+\lambda_\xi\Phi_\xi(\tau,\nu)) \frac{{}_\xi Q_\xi^{1/2}}{\partial_\xi Q_\xi/\partial\tau}(\tau,\nu)\right\} d\tau$$

$$+ O\left(T^3 \nu^{-1-4\sigma_1/5}\right) = \frac{\lambda i}{4\pi\nu} \int_{\mathcal{C}_\xi(\nu)} \frac{c_\xi(-\tau)}{|c_\xi(\tau)|} h_1(\log{}_\xi Q_\xi(\tau,\nu)) e(-\lambda\tau+\lambda_\xi\Phi_\xi(\tau,\nu)) \frac{d\tau}{{}_\xi Q_\xi^{1/2}(\tau,\nu)}$$

$$+ O\left(T^3\left[\nu^{-1-4\sigma_1/5} + \nu^{-6/5} \int_{1/\nu}^{\nu^{-3/5}} (\nu\tau)^{-2\sigma_1} d\tau\right]\right) . \qquad (10.14)$$

The second integral in (10.14) differs from the integral on the left of (10.12) only by having $\pm\hat{h}_1(\varrho)$ in place of $\hat{h}_o(\varrho)$. On noting that $h_1(0) = 0$ we can therefore repeat the process leading from (10.12) to (10.14). If $h_2(s)$ is given by the right hand side of (10.13) with h_1 in place of h_o one easily shows again that

$$\hat{h}_2(\varrho)(e^{\varrho/2}-e^{-\varrho/2})^{-1} \ll e^{-|\varrho|\sigma_o} T^2$$

for $0 < \sigma_o < 1$. Together with (10.14) and the estimates following (10.12) this yields

$$ {}_{\xi}H_{\xi}(\lambda,\lambda,\nu) << (\frac{T}{\nu})^2(\nu^{-4\sigma_o/5}+\int_{1/\nu}^{\nu^{-3/5}}(\nu\tau)^{-2\sigma_o}d\tau)+\frac{T^3}{\nu}(\nu^{-4\sigma_1/5}+\nu^{-6/5})<<T^3\nu^{-\sigma_2} $$

with a $\sigma_2>2$ if $\sigma_o>0$ and $\sigma_1>5/4$. Thus the second part of (10.11) holds.

By (6.42) and our bound for $\hat{h}_o$ the part of ${}^{\delta}_{\eta}\mathfrak{G}^{\delta'}_{\eta}$ on which $\left|\log|{}_{\eta}q_{\eta}(\tau+\delta\pi i,\nu)|\right|\geqslant 1$ contributes $<< T^3$ to ${}_{\eta}h_{\eta}(\lambda,\lambda,\nu)$. On the rest of ${}^{\delta}_{\eta}\mathfrak{G}^{\delta'}_{\eta}$ we can integrate by parts as above since

$$ \frac{\partial}{\partial\tau}\,{}_{\eta}\varphi_{\eta}(\tau,\nu) = \frac{-1}{{}_{\eta}q_{\eta}(\tau,\nu)} \quad\text{and}\quad \frac{\partial}{\partial\tau}\,{}_{\eta}\varphi_{\eta}(\tau,\nu) = -\mathrm{ch}\tau\,\sin\nu $$

by (6.14), (6.41) and (6.42). Already after one partial integration our bound for $\hat{h}_1$ yields

$$ {}_{\eta}h_{\eta}(\lambda,\lambda,\nu) << T^3 . $$

Since ${}_{\xi}h_{\xi}$ is zero except for $\xi=\eta$, (10.11) and consequently also Theorem 2 now follow.

<u>Remark</u>: In Theorem 2 the coefficient of T^2 vanishes by (2.4), (2.7) and (7.10) precisely if $\xi=(\eta_1,\eta_2,-1)$, $m=0$ and $M(\eta_1)=\eta_2$ for an M in Γ . There is an explanation for this. We note that under the above conditions

$$ e_j(M_{\xi}^{-1}(z)) = e_j(MM_{\xi}^{-1}(z)) = e_j(M_{\xi}^{-1}M_{\xi}M_{\xi^*}^{-1}(z)) = e_j(M_{\xi}^{-1}R_{\vartheta}\exp(-\varrho_{\xi}X_{\xi})(z)) , \quad j\geqslant 0 , $$

by (5.4) and the definition of ϱ_{η} in Proposition 1. Thus we obtain from (4.11), (4.14) and Lemma 5

$$ \sum_{n\in Z}\alpha_{j\xi}(n)U_{\xi}\left(z,s_j,\frac{n}{\lambda_{\xi}}\right) = -e\left(\frac{n}{\lambda_{\xi}}\varrho_{\xi}\right)\sum_{n\in Z}\alpha_{j\xi}(n)U_{\xi}\left(z,s_j,-\frac{n}{\lambda_{\xi}}\right) , $$

whence $\alpha_{j\xi}(o)=0$ for $j\geqslant 0$. In the same way we infer from (7.17) that $\alpha_{\iota\xi}(s,o)=0$ for $\iota=1,\dots,\kappa$. In other words, the left hand side of Theorem 2 vanishes identically in the case under consideration.

11. Sum formulae (second form)

In principle Theorem 1 could also be used to study the asymptotic behaviour of the generalized Kloosterman sums provided that we were able to invert the integral transforms $h \longmapsto {}_\xi h_\chi$ and $h \longmapsto {}_\xi H_\chi$. Examples [13] show that the exact solution of this problem is not always as simple as one might expect. Indeed the Fourier coefficients of holomorphic cusp forms [3] had to be brought into the picture in order to obtain neat formulae for general weight functions. A closer look, however, shows that the contribution from these holomorphic cusp forms is negligable in many applications. In this section we develop an alternative approach which is based on our functional equations but does not depend on the holomorphic cusp forms. Roughly speaking we deduce from Lemma 9 and Proposition 3 that there are simple approximate inversions of the above integral transforms. As an application we then show how these approximate inversions yield results on the asymptotic distribution of the double cosets in ${}_\xi\Gamma_\chi$.

For the following we introduce

$$ {}_\xi Z_\chi(s,m,n) = \frac{1}{\lambda_\xi \lambda_\chi} \sum_{\delta,\delta'=0}^{1} \varepsilon^\delta \varepsilon^{\delta'} \sum_\nu {}^\delta_\xi S^{\delta'}_\chi(m,n,\nu)\,\nu^{-2s} \quad , \quad \sigma > 1 \tag{11.1} $$

and

$$ M_w(s) = 2\int_0^\infty w(\varrho)\,\varrho^{2s-1}\,d\varrho \quad , \quad \sigma > 0 \; , \tag{11.2} $$

where w denotes a C^∞-function of compact support on $\mathbb{R}$.

Theorem 3. Let $\sigma_o < 1/2$ be such that $0 < \sigma_o < \min_{s_j<1}(1-s_j)$.

(i) If $\lambda = \frac{m}{\lambda_\xi}$, $\lambda' = \frac{n}{\lambda_\chi}$ and

$$ {}_\xi W_\chi(s,\lambda,\lambda') = M_w(s)\,B(1/2,s-1/2)\,\gamma_\xi(s,\lambda) + M_w(1-s)\,B(1/2,1/2-s)\,\gamma_\chi(1-s,\lambda') $$

then

$$\frac{1}{\lambda_\xi \lambda_\chi} \sum_{\delta,\delta'=0}^{1} \varepsilon^{\delta}\varepsilon'^{\delta'} \sum_{\nu} {}_{\xi}^{\delta}S_{\chi}^{\delta'}(m,n,\nu)\mathcal{W}(\nu) = \bar{\alpha}_{o\xi}(m)\,\alpha_{o\chi}(n)\,\pi M_{\mathcal{W}}(1)$$

$$+ \sum_{1/2<s_j<1} \bar{\alpha}_{j\xi}(m)\,\alpha_{j\chi}(n)\,{}_{\xi}\mathcal{W}_{\chi}(s_j,\lambda,\lambda')\,\gamma_{\xi}(1-s_j,\lambda) + \sum_{t_j>0} \bar{\alpha}_{j\xi}(m)\,\alpha_{j\chi}(n)\,{}_{\xi}\mathcal{W}_{\chi}(s_j,\lambda,\lambda')$$

$$+ \frac{1}{4\pi i} \int_{(1/2)} \sum_{\iota=1}^{\kappa} \bar{\alpha}_{\iota\xi}(s,m)\,\alpha_{\iota\chi}(s,n)\,{}_{\xi}\mathcal{W}_{\chi}(s,\lambda,\lambda')\,ds + \frac{1}{2\pi i} \int_{(\sigma_o)} M_{\mathcal{W}}(s)\{{}_{\xi}Z_{\chi}(s,m,n)$$

$$- \gamma_{\xi}(s,\lambda)\,B(1/2,s-1/2) \sum_{\iota=1}^{\kappa} \alpha_{\iota\xi}(1-s,-m)\,\alpha_{\iota\chi}(s,n)\}ds\ ,$$

unless $\xi = \vartheta$, $m = 0$ or $\chi = \vartheta'$, $n = 0$.

(ii) On the other hand we have

$$\frac{1}{\lambda_\chi} \sum_{\delta'=0}^{1} \varepsilon'^{\delta'} \sum_{\nu} {}_{\vartheta}^{o}S_{\chi}^{\delta'}(o,n,\nu)\mathcal{W}(\nu) = \sum_{1/2<s_j\le 1} \bar{\alpha}_{j\vartheta}(o)\,\alpha_{j\chi}(n)\,M_{\mathcal{W}}(s_j)\,B(1/2,s_j-1/2)$$

$$+ \frac{1}{2\pi i} \int_{(1/2)} \sum_{\iota=1}^{\kappa} \delta(\vartheta,\vartheta_\iota)\,\alpha_{\iota\chi}(s,n)\,M_{\mathcal{W}}(s)\,ds + \frac{1}{2\pi i} \int_{(\sigma_o)} M_{\mathcal{W}}(s) \Big\{ {}_{\vartheta}Z_{\chi}(s,o,n) - \frac{{}_{\vartheta}P_{\chi}(s,o,n)}{B(1/2,s)} \Big\} ds\ .$$

<u>Proof</u>: It follows from (11.1), (11.2) and Mellin's inversion formula that

$$\frac{1}{\lambda_\xi \lambda_\chi} \sum_{\delta,\delta'=0}^{1} \varepsilon^{\delta}\varepsilon'^{\delta'} \sum_{\nu} {}_{\xi}^{\delta}S_{\chi}^{\delta'}(m,n,\nu)\mathcal{W}(\nu) = \frac{1}{2\pi i} \int_{(3/2)} {}_{\xi}Z_{\chi}(s,m,n)\,M_{\mathcal{W}}(s)\,ds\ . \tag{11.3}$$

Lemma 9 and the absolute convergence of (7.12) in $\sigma > 1$ reveal that

$${}_{\xi}Z_{\chi}(s,m,n) - {}_{\xi}P_{\chi}(s,m,n)/\beta_{\xi}\lambda_{\xi}B(1/2,s)$$

is analytic and uniformly bounded by

$$\ll |s|^{1/2} \sum_{\delta,\delta'=0}^{1} \sum_{\nu} {}_{\xi}^{\delta}S_{\chi}^{\delta'}(o,o,\nu)\,\nu^{-2\sigma}\Big(\frac{1}{\nu^2} + \frac{|s|}{\nu^4+|s|}\Big) \ll |s| \tag{11.4}$$

in $\sigma_o < \sigma < 3/2$. Repeated integrations by parts in (11.2) show that $M_{w}(s)$ decays faster than any inverse power of $|s|$ in $\sigma \geq \sigma_o$. With $\mathcal{C}_-$ as in the proof of Proposition 2(ii) we thus obtain from Proposition 3(ii) and Lemma 15

$$\frac{1}{2\pi i}\int_{(3/2)} {}_{\xi}Z_{\chi}(s,m,n)M_{w}(s)\,ds = \frac{1}{2\pi i}\int_{(\sigma_o)}\left\{{}_{\xi}Z_{\chi}(s,m,n) - \frac{{}_{\xi}P_{\chi}(s,m,n)}{\beta_{\xi}\lambda_{\xi}B(1/2,s)}\right\} M_{w}(s)\,ds$$

$$+ \frac{1}{2\pi i\beta_{\xi}\lambda_{\xi}}\int_{\mathcal{C}_-} {}_{\xi}P_{\chi}(s,m,n)\,\frac{M_{w}(s)}{B(1/2,s)}\,ds + \sum_{1/2<s_j\leq 1} \bar{\alpha}_{j\xi}(m)\,\alpha_{j\chi}(n)\,M_{w}(s_j)\,B(1/2,s_j-1/2)$$

$$+\sum_{t_j\geq 0} \bar{\alpha}_{j\xi}(m)\,\alpha_{j\chi}(n)\,{}_{\xi}\mathcal{W}_{\chi}(s_j,\lambda,\lambda') + \frac{1}{2}\sum_{\iota=1}^{\kappa} \bar{\alpha}_{\iota\xi}(1/2,m)\,\alpha_{\iota\chi}(1/2,n)\,M_{w}(1/2) \quad . \qquad (11.5)$$

For the residue of

$$\left\{\frac{1}{(s-1/2)^2} + \frac{1}{2s-1}\left[\frac{\partial\gamma_{\xi}}{\partial s}(1/2,\lambda) + \frac{\partial\gamma_{\chi}}{\partial s}(1/2,\lambda')\right]\right\}\frac{2\pi M_{w}(s)}{B(1/2,s)}$$

at $s = 1/2$ equals

$$\frac{d}{ds}\left\{M_{w}(s)\,\frac{\pi\gamma_{\xi}(s,\lambda)}{B(1/2,s)} - M_{w}(1-s)\,\frac{\pi\gamma_{\chi}(1-s,\lambda')}{B(1/2,1-s)}\right\}_{s=1/2} = {}_{\xi}\mathcal{W}_{\chi}(1/2,\lambda,\lambda') \ ,$$

unless $\xi = \vartheta$, $\lambda = 0$ or $\chi = \vartheta'$, $\lambda' = 0$.

If $\xi = \vartheta$, $m = 0$ then the last two terms in (11.5) vanish as we observed in Lemma 5 and in the proof of Proposition 3 (ii) . In this case we can also shift the integration over $\mathcal{C}_-$ back to the 1/2-line. Thus Theorem 3 (ii) follows from (11.3) and (11.5) in view of (4.19) , (5.2) and (7.18) .

On the other hand if neither $\xi = \vartheta$, $m = 0$ nor $\chi = \vartheta'$, $n = 0$ Lemma 4, Proposition 3, Lemma 15 and the definition of σ_o yield

$$\int_{\mathcal{C}_-} {}_\xi P_\chi(s,m,n)\,\frac{M_{\mathcal{W}}(s)}{B(1/2,s)}\,ds = \int_{(\sigma_o)} \Big\{ \frac{\gamma_\chi(s,\lambda')}{\gamma_\xi(1-s,\lambda)}\,{}_\xi P_\chi(1-s,m,n) - \operatorname{ctg}\pi s\,{}_\xi p_\chi(s,m,n)$$

$$- \frac{\operatorname{ctg}\pi s}{\gamma_\xi(1-s,\lambda)}\,[\delta(\xi,\chi)\delta_{mn} + \delta^*(\xi,\chi)\delta_{-mn}\varepsilon e(\lambda\varrho_\xi)]\Big\}\frac{M_{\mathcal{W}}(s)}{B(1/2,s)}\,ds +$$

$$2\pi i\beta_\xi\lambda_\xi \sum_{1/2<s_j<1} \bar{\alpha}_{j\xi}(m)\alpha_{j\chi}(n)M_{\mathcal{W}}(1-s_j)B(1/2,1/2-s_j)\,\frac{\gamma_\chi(1-s_j,\lambda')}{\gamma_\xi(s_j,\lambda)} +$$

$$\beta_\xi\lambda_\xi\int_{\mathcal{C}_-} \sum_{\iota=1}^{\kappa} \alpha_{\iota\xi}(1-s,-m)\alpha_{\iota\chi}(s,n)M_{\mathcal{W}}(s)B(1/2,s-1/2)\gamma_\xi(s,\lambda)\,ds\,. \tag{11.6}$$

By definition of $\mathcal{C}_\pm$ and the first functional equation in Proposition 3(iv) the last integral equals

$$\frac{1}{2}\int_{\mathcal{C}_-} \sum_{\iota=1}^{\kappa} \alpha_{\iota\xi}(1-s,-m)\alpha_{\iota\chi}(s,n)M_{\mathcal{W}}(s)B(1/2,s-1/2)\gamma_\xi(s,\lambda)\,ds +$$

$$\frac{1}{2}\int_{\mathcal{C}_+} \sum_{\iota=1}^{\kappa} \alpha_{\iota\xi}(1-s,-m)\alpha_{\iota\chi}(s,n)M_{\mathcal{W}}(1-s)B(1/2,1/2-s)\gamma_\chi(1-s,\lambda')\,ds =$$

$$\frac{1}{2}\int_{\mathcal{C}_-} \sum_{\iota=1}^{\kappa} \alpha_{\iota\xi}(1-s,-m)\alpha_{\iota\chi}(s,n)\,{}_\xi\mathcal{W}_\chi(s,\lambda,\lambda')\,ds - \pi i M_{\mathcal{W}}(1/2)\sum_{\iota=1}^{\kappa}\alpha_{\iota\xi}(1/2,-m)\alpha_{\iota\chi}(1/2,n).$$

On the right hand side of the preceding equation we may shift the integration to the 1/2-line since ${}_\xi\mathcal{W}_\chi(s,\lambda,\lambda')$ is analytic on $\sigma = 1/2$. Thus Theorem 3 (i) follows from (11.3), (11.5) and (11.6) by a final appeal to Proposition 3(iii).

Although the last term in the identies of Theorem 3 has a rather complex appearance we demonstrate their usefulness by proving

<u>Theorem 4</u>. Suppose that $\delta = 0$ unless $\xi = \eta$ and $\delta' = 0$ unless $\chi = \eta'$. Then

$$\frac{1}{\lambda_\xi\lambda_\chi}\sum_{\nu\le X} {}^{\delta}_{\xi}S^{\delta'}_{\chi}(m,n,\nu) = \frac{\delta_{om}\delta_{on}X^2}{\beta_\xi\beta_\chi\pi\omega(\mathfrak{F})} + 2\sum_{\frac{1}{2}<s_j<1} X^{2s_j}B(3/2,s_j-1/2)\sum_{\substack{\xi\in\underline{\xi}\\ \chi\in\underline{\chi}}}$$

$$\frac{\varepsilon^{\delta}\varepsilon'^{\delta'}}{\#\underline{\xi}\#\underline{\chi}}\,\bar{\alpha}_{j\xi}(m)\alpha_{j\chi}(n) + O(X^{4/3}),\quad X\to\infty,$$

where

$$\underline{\xi} = \begin{cases} \{\xi\}\ , \text{ if } \xi = \zeta,\vartheta\ , \\ \{(\eta_1,\eta_2,\pm 1)\},\text{ if } \xi = (\eta_1,\eta_2,\varepsilon)\ , \end{cases}$$

and $\underline{\chi}$ is defined correspondingly.

Proof: First we note that

$$_{\xi}^{\delta}S_{\chi}^{\delta'}(m,n,\nu) = \frac{1}{\#\underline{\xi}\#\underline{\chi}} \sum_{\substack{\xi\in\underline{\xi}\\ \chi\in\underline{\chi}}} \varepsilon^{\delta}\varepsilon'^{\delta'} \sum_{\gamma,\gamma'=0}^{1} \varepsilon^{\gamma}\varepsilon'^{\gamma'}\ {}_{\xi}^{\gamma}S_{\chi}^{\gamma'}(m,n,\nu)\ , \tag{11.7}$$

since by Lemma 1(i) and (5.10) the last term on the right of (11.7) is the same for all ξ in $\underline{\xi}$ and χ in $\underline{\chi}$.

Next we apply Theorem 3 with functions w such that

$$w(\varrho) = \begin{cases} 1\ , & \text{if } |\varrho| \leqslant X(1-1/U)\ , \\ 0\ , & \text{if } |\varrho| \geqslant X(1+1/U)\ , \end{cases}$$

and $w^{(j)}(\varrho) \ll (\frac{U}{X})^j$

uniformly in ϱ for $j = 0,1,2,\ldots$. Here $U \geqslant 2$ is a parameter to be chosen later on. For such w's it follows from (11.2) that

$$M_w(s) = 2\int_0^X \varrho^{2s-1}d\varrho + O\left(\int_{X(1-1/U)}^{X(1+1/U)} \varrho^{2\sigma-1}d\varrho\right) = \frac{X^{2s}}{s} + O\left(\frac{X^{2\sigma}}{U}\right)$$

uniformly in $\frac{1}{2} \leqslant \sigma \leqslant 1$. On the other hand, partial integrations yield

$$M_w(s) = \frac{(-1)^j}{s(2s+1)\ldots(2s+j-1)} \int_{X(1-1/U)}^{X(1+1/U)} w^{(j)}(\varrho)\varrho^{2s+j-1}d\varrho \qquad , j = 1,2,\ldots .$$

Therefore M_w is entire except for a simple pole at $s = 0$ with residue 1 and

$$M_w(s) \ll \frac{X^{2\sigma}}{|s|}\left(\frac{U}{U+|s|}\right)^2 \quad , \quad (sM_w(s))^{(\ell)} \ll X^{2\sigma}(\log X)^{\ell}$$

uniformly in $-1 \leqslant \sigma \leqslant 2$ for $\ell = 1,2,\ldots$. Thus we conclude from Lemma 4 that

$$_{\xi}W_{\chi}(s,\lambda,\lambda')\gamma_{\xi}(1-s,\lambda) = 2B(3/2,s-1/2)X^{2s} + O\left(\frac{X^{2\sigma}}{U} + X\right)$$

for $\frac{1}{2} < s \leqslant 1$ and that on $\sigma = 1/2$

$$ {}_{\xi}\mathcal{W}_{\chi}(s,\lambda,\lambda') \ll \begin{cases} X\log X\,, & \text{if } |s|\leq 1\,,\\ X|s|^{-1/2}, & \text{otherwise}\,. \end{cases} $$

We are now in a position to handle the five terms on the right of the identity in Theorem 3 (i). Clearly the first two of them amount to

$$ 2\sum_{\frac{1}{2}<s_j\leq 1} X^{2s_j}B(3/2,s_j-1/2)\,\bar{\alpha}_{j\xi}(m)\,\alpha_{j\chi}(n) + O\left(\frac{X^2}{U}+X\right)\,, \tag{11.8} $$

while by Cauchy's inequality, partial integration and Theorem 2 the third and fourth are bounded by

$$ X\sum_{t_j>0}|\alpha_{j\xi}(m)\,\alpha_{j\chi}(n)|\,t_j^{-3/2}\left(\frac{U}{U+t_j}\right)^2 + \frac{X}{4\pi i}\int_{(1/2)}\sum_{\iota=1}^{\kappa}|\alpha_{\iota\xi}(s,m)\,\alpha_{\iota\chi}(s,n)|\,\cdot $$

$$ \cdot\left(\frac{U}{U+|s|}\right)^2\frac{ds}{|s|^{3/2}} + O(X\log X) \ll X(U^{1/2}+\log X)\,. \tag{11.9} $$

In the fifth we substitute the sum involving the Fourier coefficients of the Eisenstein series by means of Proposition 3(iii). Since by Lemma 11 and (7.11)

$$ \gamma_{\chi}(s,\lambda')\,{}_{\xi}p_{\chi}(1-s,m,n) \ll 1 $$

in $0<\sigma<1$ this term is

$$ \ll\left|\int_{(\sigma_o)}M_{\mathcal{W}}(s)\left\{{}_{\xi}Z_{\chi}(s,m,n) - \frac{{}_{\xi}P_{\chi}(s,m,n)}{\beta_{\xi}\lambda_{\xi}B(1/2,s)} + O(|s|^{1/2})\right\}ds\right| $$

$$ + \left|\int_{(\sigma_o)}M_{\mathcal{W}}(s)\frac{\gamma_{\xi}(s,\lambda)\,\gamma_{\chi}(s,\lambda')}{B(1/2,s)}\,{}_{\xi}P_{\chi}(1-s,m,n)\,ds\right| \ll X^{2\sigma_o}U\,. \tag{11.10} $$

For the first integral in (11.10) is

$$ X^{2\sigma_o}\int_0^{\infty}\left(\frac{U}{U+t}\right)^2dt \ll X^{2\sigma_o}U $$

by (11.4) and our estimates for $M_{\mathcal{W}}(s)$. On the other hand, by Lemma 4 and Proposition 3(ii) the second integrand in (11.10) is analytic in the strip $-\sigma_o\leq\sigma\leq\sigma_o$ except possibly for a pole of order ≤ 3 at

$s = 0$. Thus by Cauchy's theorem and our estimates for $M_w(s)$ the second integral in (11.10) is

$$<< \log^3 X + X^{-2\sigma_o} \int_0^\infty t^{-1/2} \left(\frac{U}{U+t}\right)^2 dt << X^{2\sigma_o} U$$

since ${}_\xi P_\chi(1-s,m,n)$ is bounded on $\sigma = -\sigma_o$.

The three terms on the right of the identity in Theorem 3(ii) can be treated very similarly. Indeed the first equals (11.8) in case $\xi = \vartheta$, $m = 0$ while, again by Theorem 2 and (11.4), the second and third are bounded by the right hand side of (11.9) and (11.10) respectively.

It is clear that the characteristic function of the interval $(-X,X)$ is majorized or minorized by some of the weight functions under consideration. Hence Theorem 3, (11.7)-(11.10) and a simple Tauberian argument yield the expression

$$2 \sum_{1/2 \leq s_j \leq 1} X^{2s_j} B(3/2, s_j - 1/2) \sum_{\substack{\xi \in \underline{\xi} \\ \chi \in \underline{\chi}}} \frac{\varepsilon^\delta \varepsilon'^{\delta'}}{\#\underline{\xi} \#\underline{\chi}} \alpha_{j\xi}(m) \alpha_{j\chi}(n) +$$

$$O\left(\frac{X^2}{U} + XU^{1/2} + X \log X + X^{2\sigma_o} U\right)$$

for the left hand side of the equation we set out to prove. Lemma 5 shows that $s_o = 1$ contributes to the preceding sum precisely

$$\frac{\delta_{om} \delta_{on} X^2}{\beta_\xi \beta_\chi \pi \omega(\mathfrak{F})} .$$

Thus Theorem 4 follows from choosing $U = X^{2/3}$ and $\sigma_o \leq 1/3$.

An immediate consequence of (5.7), (5.10), Theorem 4 and Weyl's criterion on uniform distribution is the following

Corollary. Let $0 \leq a_1 < a_2 \leq \lambda_\xi$, $0 \leq b_1 < b_2 \leq \lambda_\chi$. Suppose that $\delta = 0$ unless $\xi = \eta$ and $\delta' = 0$ unless $\chi = \eta'$. Then the number of double cosets $\Gamma_\xi M_\xi^{-1} M M_\chi \Gamma_\chi$ in Γ such that

$$a_1 \leq {}_\xi\Lambda^{\ell}_\chi(M) \leq a_2 \pmod{\lambda_\xi} , \qquad b_1 \leq {}_\xi\Lambda^{r}_\chi(M) \leq b_2 \pmod{\lambda_\chi} ,$$

$${}_\xi\delta_\chi(M) = \delta , \qquad {}_\xi\delta'_\chi(M) = \delta' \quad \text{and} \quad {}_\xi\nu_\chi(M) \leq X$$

is asymptotically equal to

$$\frac{(a_2-a_1)(b_2-b_1)}{\beta_\xi\beta_\chi\pi\omega(\mathfrak{F})} X^2 , \quad \text{as} \quad X \to \infty .$$

To give the reader an idea of what is subsumed in Theorem 4 we end by listing a few examples.

Example 1. If $d(\cdot,\cdot)$ denotes the hyperbolic distance function then by Lemma 1 and (5.1)

$$d(\zeta, M(\zeta')) = d(i, M_\zeta M M_{\zeta'}^{-1}(i)) = d(i, i\, \nu^2_{\zeta'}(M_\zeta M M_{\zeta'}^{-1})) .$$

Since

$$d(i, iy) = \log y$$

for $y \geq 1$, we infer from Lemma 1, (5.2), (5.7) and (5.10) that

$$\frac{4\pi^2}{\lambda_\zeta\lambda_{\zeta'}} \sum_{\nu \leq X} {}_\zeta^{o}S^{o}_{\zeta'}(o,o,\nu)$$

is just the number of points $M(\zeta')$ in $\mathfrak{H}$ such that M is in Γ and $0 < d(\zeta, M(\zeta')) \leq 2 \log X$. In other words, if $\xi = \zeta$, $m = 0$ and $\chi = \zeta'$, $n = 0$, Theorem 4 yields the usual bound for the lattice point problem in hyperbolic circles. A different proof of Theorem 4 for $\xi = \zeta$, $\chi = \zeta'$ and arbitrary m,n was discussed by A. Selberg in a lecture at Rockefeller University in 1977.

Example 2. Under the assumptions and in the notation of Corollary 1 to Theorem 1 we infer from Theorem 4 for $mn \neq 0$

$$\sum_{0<c\leq X} S(m,n,c) = \sum_{1/2<s_j<1} \bar{\alpha}_{jm}\alpha_{jn} \frac{\Gamma(2s_j-1)\quad X^{2s_j}}{(2\pi)^{2s_j-1}\,|mn|^{s_j-1/2}\quad 2s_j} + O(X^{4/3}) , \quad X \to \infty .$$

This estimate also follows from the results in [18]. If Γ is a congruence subgroup of $SL_2(Z)$ the error term can even be improved to $O(X^{\alpha})$ for every $\alpha > 7/6$ ([3],[13]). This is due to the fact that $S(m,n,c)$ is then expressible as a sum over residues mod c (cf. Remark 1 to Lemma 6). Therefore optimal bounds for the individual $S(m,n,c)$ are available by A. Weil's estimates for such exponential sums. If corresponding bounds also held for other cases similar progress could then be made there.

Example 3. If $\eta = (\eta_1, \eta_2, \varepsilon)$ is such that $-\infty < \eta_1 < \eta_2 < \infty$ let L_o denote the oriented geodesic in $\mathfrak{H}$ leading from η_1 to η_2. By (5.1) we may choose

$$M_\eta = (\eta_2 - \eta_1)^{-1/2} \begin{pmatrix} 1 & -\eta_1 \\ -1 & \eta_2 \end{pmatrix} .$$

We say that two oriented geodesics in $\mathfrak{H}$ are in the same Γ- or η-class if they can be transformed into each other by a matrix in Γ or Γ_η respectively. Thus there is a 1-1 correspondence between the double cosets $\Gamma_\eta M \Gamma_\eta$, M in Γ, and the η-classes in the Γ-class of L_o. In order to express part of the last corollary in terms of η-class invariants we note that an oriented geodesic L is uniquely determined by its starting point $\mu_1 = \mu_1(L)$ and its endpoint $\mu_2 = \mu_2(L)$. Suppose now that $L = M(L_o)$, i.e. $\mu_j = M(\eta_j)$ for $j = 1,2$. Then we infer from (3.4), (3.7), (3.13), (5.1) and (5.2) that

$${}_\eta\Lambda^{\ell}_\eta(M_\eta M M_\eta^{-1}) = \frac{1}{2} \log \left| \frac{(\mu_1 - \eta_1)(\mu_2 - \eta_1)}{(\mu_1 - \eta_2)(\mu_2 - \eta_2)} \right| = n\lambda_\eta + \{L\}_\eta ,$$

say, where the integer n is so chosen that $\{L\}_\eta$ lies in the interval $[0, \lambda_\eta)$. Moreover, if $M' = M_\eta M M_\eta^{-1}$ lies in ${}_\eta G_\eta$ and Lemma 1(i) applies with $\delta = 0$ and $\delta' = 1$ we obtain for $\nu = {}_\eta\nu_\eta(M')$

$$\frac{M'(o)}{M'(\infty)} = \left(\frac{\nu^2 - 1}{\nu^2 + 1}\right)^2 = \frac{\mu_1 - \eta_1}{\mu_1 - \eta_2} \Big/ \frac{\mu_2 - \eta_1}{\mu_2 - \eta_2} = D_\eta(L) ,$$

say, where the cross ratio $D_\eta(L)$ lies between 0 and 1. Note also that $\delta = 0$ and $\delta' = 1$ are equivalent with $\eta_1 < \mu_1 < \mu_2 < \eta_2$. These inequalities, $\{L\}_\eta$ and $D_\eta(L)$ only depend on the η-class of L. Therefore no confusion should arise if we subsequently use the same notation for a geodesic and its η-class. The inversion in Γ induces an involution $L \mapsto L^\iota$ on η- classes. Thus by (3.15) we have

$$-\Lambda^r_\eta(M_\eta M M_\eta^{-1}) = n\lambda_\eta + \{L^\iota\}_\eta$$

with a suitable integer n.

In terms of η-class invariants the last Corollary now reads in case of $\xi = \chi = \eta$, $\delta = 0$ and $\delta' = 1$: The number of η-classes L in the Γ-class of L_o such that $a_1 \leqslant \{L\}_\eta \leqslant a_2$, $b_1 \leqslant \{L^\iota\}_\eta \leqslant b_2$, $\eta_1 < \mu_1(L) < \mu_2(L) < \eta_2$ and $D_\eta(L) \leqslant \varrho$ is asymptotically equal to

$$\frac{(a_2-a_1)(b_2-b_1)}{4\pi\omega(\mathfrak{F})}\left(\frac{1+\varrho^{1/2}}{1-\varrho^{1/2}}\right), \quad \varrho \uparrow 1 .$$

By a familiar transition this example can also be formulated as a counting problem for indefinite binary quadratic forms if $\Gamma = SL_2(\mathbb{Z})$.

Literature

[1] Bruggeman, R.W.: Fourier coefficients of cusp forms. Invent. Math. 45, 1-18(1978).

[2] Bruggeman, R.W.: Fourier coefficients of automorphic forms. Lecture Notes in Math. 865. Berlin-Heidelberg-New York: Springer 1981.

[3] Deshouillers, J.M., Iwaniec, H.: Kloosterman sums and Fourier coefficients of cusp forms. Invent. Math. 70, 219-288(1982).

[4] Deshouillers, J.M., Iwaniec, H.: Power mean values for the Riemann zeta-function. Mathematika 29, 202-212 (1982).

[5] Fay, J.D.: Fourier coefficients of the resolvent for a Fuchsian group. J. Reine Angew. Math. 293-294, 143-203 (1977).

[6] Goldfeld, D., Sarnak. P.: Sums of Kloosterman sums. Invent. Math. 71, 243-250(1983).

[7] Good, A.: Cusp forms and eigenfunctions of the Laplacian. Math. Ann. 255, 523-548(1981).

[8] Good, A.: The square mean of Dirichlet series associated to cusp forms. Mathematika 29, 278-295(1982).

[9] Good, A.: On various means involving the Fourier coefficients of cusp forms. Math. Zeitschrift 183, 95-129(1983).

[10] Heijal, D.H.: Sur quelques propriétés asymptotiques des périodes hyperboliques et des invariants algébriques d'un sous-groupe discret de PSL(2,R). Comptes rendues 294, Sér. I, 509-512(1982).

[11] Iwaniec, H.: Fourier coefficients of cusp forms and the Riemann zeta-function. Sém. Th.Nb. Bordeaux (1979-1980), exposé 18,36 p..

[12] Kubota, T.: Elementary theory of Eisenstein series. New York: Wiley 1973.

[13] Kuznietsov, N.V.: Petersson hypothesis for parabolic forms of weight zero and Linnik hypothesis. Sums of Kloosterman sums. Math. Sborn. 111(153), 334-383(1980).

[14] Magnus, W., Oberhettinger, F., Soni R.P.: Formulas and theorems for the special functions of mathematical physics, 3rded. Berlin-Heidelberg-New York: Springer 1966.

[15] Neunhöffer, H.: Ueber die analytische Fortsetzung von Poincaréreihen. Sitzb. Heidelberg Akad. Wiss. (Mat.-Nat.Kl.) 2Abh., 33-90(1973).

[16] Niebur, D.: A class of nonanalytic automorphic functions. Nagoya Math. J. 52, 133-145 (1973).

[17] Petersson, H.: Einheitliche Begründung der Vollständigkeitssätze für die Poincaréschen Reihen von reeller Dimension bei beliebigen Grenzkreisgruppen erster Art. Abh. Math. Sem. Hamburg 14, 22-60 (1941).

[18] Proskurin, N.V.: Summation formulas for generalized Kloosterman sums. Zap. Naucn.Sem. Leningrad. Otdel. Math. Inst. Steklov 82, 103-135(1979).

[19] Roelcke, W.: Ueber die Wellengleichung bei Grenzkreisgruppen erster Art. Sitzb. Heidelberg Akad. Wiss. (Mat.-Nat.Kl.) 4 Abh., 159-267(1956).

[20] Selberg, A.: Harmonic analysis and discontinuous groups in weakly symmetric Riemannian spaces with applications to Dirichlet series. J. Indian Math. Soc. 20, 47-87(1956).

[21] Selberg, A.: On the estimation of Fourier coefficients of modular forms. Proc. Symp. in Pure Math. VIII, pp. 1-15. Providence: Amer. Math. Soc. 1965.

[22] Shimura, G.: Introduction to the arithmetic theory of automorphic functions. Princeton: University Press 1971.

Index of notations

Latin (capital)

Latin (lower case)

Index of terminology

Vol. 900: P. Deligne, J. S. Milne, A. Ogus, and K.-Y. Shih, Hodge Cycles, Motives, and Shimura Varieties. V, 414 pages. 1982.

Vol. 901: Séminaire Bourbaki vol. 1980/81 Exposés 561–578. III, 299 pages. 1981.

Vol. 902: F. Dumortier, P.R. Rodrigues, and R. Roussarie, Germs of Diffeomorphisms in the Plane. IV, 197 pages. 1981.

Vol. 903: Representations of Algebras. Proceedings, 1980. Edited by M. Auslander and E. Lluis. XV, 371 pages. 1981.

Vol. 904: K. Donner, Extension of Positive Operators and Korovkin Theorems. XII, 182 pages. 1982.

Vol. 905: Differential Geometric Methods in Mathematical Physics. Proceedings, 1980. Edited by H.-D. Doebner, S.J. Andersson, and H.R. Petry. VI, 309 pages. 1982.

Vol. 906: Séminaire de Théorie du Potentiel, Paris, No. 6. Proceedings. Edité par F. Hirsch et G. Mokobodzki. IV, 328 pages. 1982.

Vol. 907: P. Schenzel, Dualisierende Komplexe in der lokalen Algebra und Buchsbaum-Ringe. VII, 161 Seiten. 1982.

Vol. 908: Harmonic Analysis. Proceedings, 1981. Edited by F. Ricci and G. Weiss. V, 325 pages. 1982.

Vol. 909: Numerical Analysis. Proceedings, 1981. Edited by J.P. Hennart. VII, 247 pages. 1982.

Vol. 910: S.S. Abhyankar, Weighted Expansions for Canonical Desingularization. VII, 236 pages. 1982.

Vol. 911: O.G. Jørsboe, L. Mejlbro, The Carleson-Hunt Theorem on Fourier Series. IV, 123 pages. 1982.

Vol. 912: Numerical Analysis. Proceedings, 1981. Edited by G. A. Watson. XIII, 245 pages. 1982.

Vol. 913: O. Tammi, Extremum Problems for Bounded Univalent Functions II. VI, 168 pages. 1982.

Vol. 914: M. L. Warshauer, The Witt Group of Degree k Maps and Asymmetric Inner Product Spaces. IV, 269 pages. 1982.

Vol. 915: Categorical Aspects of Topology and Analysis. Proceedings, 1981. Edited by B. Banaschewski. XI, 385 pages. 1982.

Vol. 916: K.-U. Grusa, Zweidimensionale, interpolierende Lg-Splines und ihre Anwendungen. VIII, 238 Seiten. 1982.

Vol. 917: Brauer Groups in Ring Theory and Algebraic Geometry. Proceedings, 1981. Edited by F. van Oystaeyen and A. Verschoren. VIII, 300 pages. 1982.

Vol. 918: Z. Semadeni, Schauder Bases in Banach Spaces of Continuous Functions. V, 136 pages. 1982.

Vol. 919: Séminaire Pierre Lelong – Henri Skoda (Analyse) Années 1980/81 et Colloque de Wimereux, Mai 1981. Proceedings. Edité par P. Lelong et H. Skoda. VII, 383 pages. 1982.

Vol. 920: Séminaire de Probabilités XVI, 1980/81. Proceedings. Edité par J. Azéma et M. Yor. V, 622 pages. 1982.

Vol. 921: Séminaire de Probabilités XVI, 1980/81. Supplément: Géométrie Différentielle Stochastique. Proceedings. Edité par J. Azéma et M. Yor. III, 285 pages. 1982.

Vol. 922: B. Dacorogna, Weak Continuity and Weak Lower Semicontinuity of Non-Linear Functionals. V, 120 pages. 1982.

Vol. 923: Functional Analysis in Markov Processes. Proceedings, 1981. Edited by M. Fukushima. V, 307 pages. 1982.

Vol. 924: Séminaire d'Algèbre Paul Dubreil et Marie-Paule Malliavin. Proceedings, 1981. Edité par M.-P. Malliavin. V, 461 pages. 1982.

Vol. 925: The Riemann Problem, Complete Integrability and Arithmetic Applications. Proceedings, 1979-1980. Edited by D. Chudnovsky and G. Chudnovsky. VI, 373 pages. 1982.

Vol. 926: Geometric Techniques in Gauge Theories. Proceedings, 1981. Edited by R. Martini and E.M.de Jager. IX, 219 pages. 1982.

Vol. 927: Y. Z. Flicker, The Trace Formula and Base Change for GL (3). XII, 204 pages. 1982.

Vol. 928: Probability Measures on Groups. Proceedings 1981. Edited by H. Heyer. X, 477 pages. 1982.

Vol. 929: Ecole d'Eté de Probabilités de Saint-Flour X – 1980. Proceedings, 1980. Edited by P.L. Hennequin. X, 313 pages. 1982.

Vol. 930: P. Berthelot, L. Breen, et W. Messing, Théorie de Dieudonné Cristalline II. XI, 261 pages. 1982.

Vol. 931: D.M. Arnold, Finite Rank Torsion Free Abelian Groups and Rings. VII, 191 pages. 1982.

Vol. 932: Analytic Theory of Continued Fractions. Proceedings, 1981. Edited by W.B. Jones, W.J. Thron, and H. Waadeland. VI, 240 pages. 1982.

Vol. 933: Lie Algebras and Related Topics. Proceedings, 1981. Edited by D. Winter. VI, 236 pages. 1982.

Vol. 934: M. Sakai, Quadrature Domains. IV, 133 pages. 1982.

Vol. 935: R. Sot, Simple Morphisms in Algebraic Geometry. IV, 146 pages. 1982.

Vol. 936: S.M. Khaleelulla, Counterexamples in Topological Vector Spaces. XXI, 179 pages. 1982.

Vol. 937: E. Combet, Intégrales Exponentielles. VIII, 114 pages. 1982.

Vol. 938: Number Theory. Proceedings, 1981. Edited by K. Alladi. IX, 177 pages. 1982.

Vol. 939: Martingale Theory in Harmonic Analysis and Banach Spaces. Proceedings, 1981. Edited by J.-A. Chao and W.A. Woyczyński. VIII, 225 pages. 1982.

Vol. 940: S. Shelah, Proper Forcing. XXIX, 496 pages. 1982.

Vol. 941: A. Legrand, Homotopie des Espaces de Sections. VII, 132 pages. 1982.

Vol. 942: Theory and Applications of Singular Perturbations. Proceedings, 1981. Edited by W. Eckhaus and E.M. de Jager. V, 363 pages. 1982.

Vol. 943: V. Ancona, G. Tomassini, Modifications Analytiques. IV, 120 pages. 1982.

Vol. 944: Representations of Algebras. Workshop Proceedings, 1980. Edited by M. Auslander and E. Lluis. V, 258 pages. 1982.

Vol. 945: Measure Theory. Oberwolfach 1981, Proceedings. Edited by D. Kölzow and D. Maharam-Stone. XV, 431 pages. 1982.

Vol. 946: N. Spaltenstein, Classes Unipotentes et Sous-groupes de Borel. IX, 259 pages. 1982.

Vol. 947: Algebraic Threefolds. Proceedings, 1981. Edited by A. Conte. VII, 315 pages. 1982.

Vol. 948: Functional Analysis. Proceedings, 1981. Edited by D. Butković, H. Kraljević, and S. Kurepa. X, 239 pages. 1982.

Vol. 949: Harmonic Maps. Proceedings, 1980. Edited by R.J. Knill, M. Kalka and H.C.J. Sealey. V, 158 pages. 1982.

Vol. 950: Complex Analysis. Proceedings, 1980. Edited by J. Eells. IV, 428 pages. 1982.

Vol. 951: Advances in Non-Commutative Ring Theory. Proceedings, 1981. Edited by P.J. Fleury. V, 142 pages. 1982.

Vol. 952: Combinatorial Mathematics IX. Proceedings, 1981. Edited by E. Billington, S. Oates-Williams, and A.P. Street. XI, 443 pages. 1982.

Vol. 953: Iterative Solution of Nonlinear Systems of Equations. Proceedings, 1982. Edited by R. Ansorge, Th. Meis, and W. Törnig. VII, 202 pages. 1982.

Vol. 954: S.G. Pandit, S.G. Deo, Differential Systems Involving Impulses. VII, 102 pages. 1982.

Vol. 955: G. Gierz, Bundles of Topological Vector Spaces and Their Duality. IV, 296 pages. 1982.

Vol. 956: Group Actions and Vector Fields. Proceedings, 1981. Edited by J.B. Carrell. V, 144 pages. 1982.

Vol. 957: Differential Equations. Proceedings, 1981. Edited by D.G. de Figueiredo. VIII, 301 pages. 1982.

Vol. 958: F.R. Beyl, J. Tappe, Group Extensions, Representations, and the Schur Multiplicator. IV, 278 pages. 1982.

Vol. 959: Géométrie Algébrique Réelle et Formes Quadratiques, Proceedings, 1981. Edité par J.-L. Colliot-Thélène, M. Coste, L. Mahé, et M.-F. Roy. X, 458 pages. 1982.

Vol. 960: Multigrid Methods. Proceedings, 1981. Edited by W. Hackbusch and U. Trottenberg. VII, 652 pages. 1982.

Vol. 961: Algebraic Geometry. Proceedings, 1981. Edited by J.M. Aroca, R. Buchweitz, M. Giusti, and M. Merle. X, 500 pages. 1982.

Vol. 962: Category Theory. Proceedings, 1981. Edited by K.H. Kamps, D. Pumplün, and W. Tholen, XV, 322 pages. 1982.

Vol. 963: R. Nottrot, Optimal Processes on Manifolds. VI, 124 pages. 1982.

Vol. 964: Ordinary and Partial Differential Equations. Proceedings, 1982. Edited by W.N. Everitt and B.D. Sleeman. XVIII, 726 pages. 1982.

Vol. 965: Topics in Numerical Analysis. Proceedings, 1981. Edited by P.R. Turner. IX, 202 pages. 1982.

Vol. 966: Algebraic K-Theory. Proceedings, 1980, Part I. Edited by R.K. Dennis. VIII, 407 pages. 1982.

Vol. 967: Algebraic K-Theory. Proceedings, 1980. Part II. VIII, 409 pages. 1982.

Vol. 968: Numerical Integration of Differential Equations and Large Linear Systems. Proceedings, 1980. Edited by J. Hinze. VI, 412 pages. 1982.

Vol. 969: Combinatorial Theory. Proceedings, 1982. Edited by D. Jungnickel and K. Vedder. V, 326 pages. 1982.

Vol. 970: Twistor Geometry and Non-Linear Systems. Proceedings, 1980. Edited by H.-D. Doebner and T.D. Palev. V, 216 pages. 1982.

Vol. 971: Kleinian Groups and Related Topics. Proceedings, 1981. Edited by D.M. Gallo and R.M. Porter. V, 117 pages. 1983.

Vol. 972: Nonlinear Filtering and Stochastic Control. Proceedings, 1981. Edited by S.K. Mitter and A. Moro. VIII, 297 pages. 1983.

Vol. 973: Matrix Pencils. Proceedings, 1982. Edited by B. Kågström and A. Ruhe. XI, 293 pages. 1983.

Vol. 974: A. Draux, Polynômes Orthogonaux Formels – Applications. VI, 625 pages. 1983.

Vol. 975: Radical Banach Algebras and Automatic Continuity. Proceedings, 1981. Edited by J.M. Bachar, W.G. Bade, P.C. Curtis Jr., H.G. Dales and M.P. Thomas. VIII, 470 pages. 1983.

Vol. 976: X. Fernique, P.W. Millar, D.W. Stroock, M. Weber, Ecole d'Eté de Probabilités de Saint-Flour XI – 1981. Edited by P.L. Hennequin. XI, 465 pages. 1983.

Vol. 977: T. Parthasarathy, On Global Univalence Theorems. VIII, 106 pages. 1983.

Vol. 978: J. Ławrynowicz, J. Krzyż, Quasiconformal Mappings in the Plane. VI, 177 pages. 1983.

Vol. 979: Mathematical Theories of Optimization. Proceedings, 1981. Edited by J.P. Cecconi and T. Zolezzi. V, 268 pages. 1983.

Vol. 980: L. Breen. Fonctions thêta et théorème du cube. XIII, 115 pages. 1983.

Vol. 981: Value Distribution Theory. Proceedings, 1981. Edited by I. Laine and S. Rickman. VIII, 245 pages. 1983.

Vol. 982: Stability Problems for Stochastic Models. Proceedings, 1982. Edited by V.V. Kalashnikov and V.M. Zolotarev. XVII, 295 pages. 1983.

Vol. 983: Nonstandard Analysis-Recent Developments. Edited by A.E. Hurd. V, 213 pages. 1983.

Vol. 984: A. Bove, J.E. Lewis, C. Parenti, Propagation of Singularities for Fuchsian Operators. IV, 161 pages. 1983.

Vol. 985: Asymptotic Analysis II. Edited by F. Verhulst. VI, 497 pages. 1983.

Vol. 986: Séminaire de Probabilités XVII 1981/82. Proceedings. Edited by J. Azéma and M. Yor. V, 512 pages. 1983.

Vol. 987: C.J. Bushnell, A. Fröhlich, Gauss Sums and p-adic Division Algebras. XI, 187 pages. 1983.

Vol. 988: J. Schwermer, Kohomologie arithmetisch definierter Gruppen und Eisensteinreihen. III, 170 pages. 1983.

Vol. 989: A.B. Mingarelli, Volterra-Stieltjes Integral Equations and Generalized Ordinary Differential Expressions. XIV, 318 pages. 1983.

Vol. 990: Probability in Banach Spaces IV. Proceedings, 1982. Edited by A. Beck and K. Jacobs. V, 234 pages. 1983.

Vol. 991: Banach Space Theory and its Applications. Proceedings, 1981. Edited by A. Pietsch, N. Popa and I. Singer. X, 302 pages. 1983.

Vol. 992: Harmonic Analysis, Proceedings, 1982. Edited by G. Mauceri, F. Ricci and G. Weiss. X, 449 pages. 1983.

Vol. 993: R.D. Bourgin, Geometric Aspects of Convex Sets with the Radon-Nikodým Property. XII, 474 pages. 1983.

Vol. 994: J.-L. Journé, Calderón-Zygmund Operators, Pseudo-Differential Operators and the Cauchy Integral of Calderón. VI, 129 pages. 1983.

Vol. 995: Banach Spaces, Harmonic Analysis, and Probability Theory. Proceedings, 1980–1981. Edited by R.C. Blei and S.J. Sidney. V, 173 pages. 1983.

Vol. 996: Invariant Theory. Proceedings, 1982. Edited by F. Gherardelli. V, 159 pages. 1983.

Vol. 997: Algebraic Geometry – Open Problems. Edited by C. Ciliberto, F. Ghione and F. Orecchia. VIII, 411 pages. 1983.

Vol. 998: Recent Developments in the Algebraic, Analytical, and Topological Theory of Semigroups. Proceedings, 1981. Edited by K.H. Hofmann, H. Jürgensen and H.J. Weinert. VI, 486 pages. 1983.

Vol. 999: C. Preston, Iterates of Maps on an Interval. VII, 205 pages. 1983.

Vol. 1000: H. Hopf, Differential Geometry in the Large, VII, 184 pages. 1983.

Vol. 1001: D.A. Hejhal, The Selberg Trace Formula for PSL(2, ℝ). Volume 2. VIII, 806 pages. 1983.

Vol. 1002: A. Edrei, E.B. Saff, R.S. Varga, Zeros of Sections of Power Series. VIII, 115 pages. 1983.

Vol. 1003: J. Schmets, Spaces of Vector-Valued Continuous Functions. VI, 117 pages. 1983.

Vol. 1004: Universal Algebra and Lattice Theory. Proceedings, 1982. Edited by R.S. Freese and O.C. Garcia. VI, 308 pages. 1983.

Vol. 1005: Numerical Methods. Proceedings, 1982. Edited by V. Pereyra and A. Reinoza. V, 296 pages. 1983.

Vol. 1006: Abelian Group Theory. Proceedings, 1982/83. Edited by R. Göbel, L. Lady and A. Mader. XVI, 771 pages. 1983.

Vol. 1007: Geometric Dynamics. Proceedings, 1981. Edited by J. Palis Jr. IX, 827 pages. 1983.

Vol. 1008: Algebraic Geometry. Proceedings, 1981. Edited by J. Dolgachev. V, 138 pages. 1983.

Vol. 1009: T.A. Chapman, Controlled Simple Homotopy Theory and Applications. III, 94 pages. 1983.

Vol. 1010: J.-E. Dies, Chaînes de Markov sur les permutations. IX, 226 pages. 1983.